Building A Bridge

from Training to Testing

Marsha Smith
&
Shalini Bosbyshell

Building A Bridge
from Training to Testing

Copyright © 2005 by Marsha Smith & Shalini Bosbyshell

All rights reserved. No part of this book may be reproduced or transmitted in any form or by any means, electronic or mechanical, including photocopying or any information storage and retrieval system, without permission in writing from the copyright holders.

First Edition

ISBN 0-615-12775-4

Printed in the United States of America by Fidlar Doubleday, Inc.

Original Cover Art by Jason Reed Smith, www.jasonreedsmith.com
Layout & Design by Shalini Bosbyshell
Photographs (Ch. 6) by Steve Surfman, www.stevesurfman.com
Original Art (Ch. 5, 7) by Marsha Smith
Graphics (Ch. 3, 4, 7) by Shalini Bosbyshell

For information please contact: info@building-a-bridge.com

www.building-a-bridge.com

Contents

Acknowledgments	vii
Authors' Notes	viii
Introduction	1
Chapter 1 - The Bridge	**8**
Three Modes of Training	8
Wow-Confidence-Testing vs. Beginner-Intermediate-Advanced	10
Chapter 2 - Defining the Three Modes of Training	**12**
Wow - The Motivational Mode	12
The Confidence Building Mode	14
The Testing Mode	15
All Three Modes	17
Chapter 3 - Pie Charts & Percentages	**19**
Chapter 4 - Training Advice & Suggestions	**23**
The "C" Word - Correction vs. Punishment	23
Uses for Food in Training	28
Back-Stepping	31
Using a Training Log	32
Chapter 5 - The Progressive Class	**34**
The Two-Part Foundation—Part I	
An Example of the Progressive Class	40
Chapter 6 - Attention Training	**53**
The Two-Part Foundation—Part II	
Stages of Attention	56
Chapter 7- Bridging the Gap	**68**
The Novice Recall Exercise:	72
in Wow	72
in Confidence Building	73
in Testing	73
The Heel Free Exercise:	76
in Wow	76
in Confidence Building	78
in Testing	79
Conclusion	82
Glossary	84
About the Authors	88

Acknowledgments

I am grateful to my parents who, after finally giving in to my pleading, made it possible for me to have my first dog Chester, the Manchester Terrier.

Heartfelt thanks to my first Utility Dog Becca, the Scottie, who showed me that it wasn't all that difficult to be a team. My appreciation and thanks to the Sheltie Kestrel, who taught me how to win with grace.

Special thanks to my family—Richard, Jason, and Dana, who share their lives with my dogs and my goals, with additional gratitude to my son Jason for our cover art. My sincere appreciation to all my students, human and canine, past and present, who have provided inspiration and guidance throughout the years.

Grateful thanks to Shalini for helping me give birth to this project. Without her suggestions, encouragement, and input—especially in light of my non-existent relationship with computers, this book could never have become a reality. And I thank her for opening my mind to a deeper understanding of communication, adding a new dimension to my training as well as to my life with my dogs.

Thanks to Chinese Crested Dante for teaching me the real meaning of patience and for helping us illustrate the description of the Attention Training exercises in Chapter 6 with his expert demonstration.

Also gracing the photos in Chapter 6 is Chinese Crested Jack, my Obedience Trial Champion. With his distinctive style, his keen intelligence, his courage, and his understanding of true collaboration, it is my Jack who has made me the trainer—and the person—I am today. To Jack, I owe my deepest gratitude and undying love.

~Marsha Smith, 2005

I am thankful to all the dogs who have been my training partners; my first dog, Pembroke Welsh Corgi Devon, who trained me, and Mulie and Cosmo, our mixed breed dogs who patiently endured my earliest attempts to learn the basic obedience exercises.

To Great Dane Bhima, who taught the *oneness of all* to anyone who had the eyes to see—my dearest, heartfelt thanks for innumerable blessings. I remain humbled by his graciousness and intellect.

I am grateful to the elegant Scottish Deerhound Mushika, who with her enthusiasm, humor, and sensitivity first taught me the finer elements of canine dressage.

Special thanks to delightful Scottish Deerhound Peri (the Divine Miss P!) who, in her acting debut, helped illustrate Stage 3 of the Attention Training exercises in Chapter 6 by *pretending* to be distracted. Her sharp wit and clever ideas make her a stimulating training partner, as well as a source of constant amazement.

I am endlessly grateful to all my clients of many species who have taught me so many invaluable lessons on so many levels.

Thank you to my good friend, photographer Steve Surfman for the photographs illustrating Chapter 6.

Loving thanks to my husband, Hal, for his devoted support throughout my work on this project and most especially for his insightful critique and exacting editing skills which have added enormously to the readability of this book.

Sincere thanks and gratitude to Marsha for teaching me so much about training, and for her infinite patience during those times when my life interrupted my contribution to this project. And most of all, I thank Marsha, for allowing me to assist her in offering people and their dogs a chance to learn about her gifts to the world of obedience training.

~Shalini Bosbyshell, 2005

Authors' Notes

Marsha

I have loved dogs all my life—in fact, my first word was "dog." As a very small child I was not allowed to have a dog of my own, and yet even then, dogs were the focus of my life. I would follow other people's dogs down the street just to be near them.

There is an image that remains in my mind... an image of happy people sledding on a snowy day, their beautiful dogs running along side, people and dogs having so much fun together, sharing in each other's lives. I watched that joyful scene from my window and quietly wished for dogs of my own; dogs who would be more than pets, dogs with whom I could share my life. It is difficult to explain the depth of that longing. It was so much more than simply wishing for a dog of my own to play with and care for. And although I can't say that as a small child I knew my life's mission, I do recall feeling that it was somehow imperative that I explore the human-dog relationship firsthand.

Today, I look back on those childhood experiences and see that I truly felt in my heart that it would be through partnership with dogs that I would fulfill my desire to help others.

As an adult, while participating in obedience trials, I began to notice something about many of the performances I was watching. If you go to any obedience competition and watch for a while, I'm sure that you'll see the same thing. Far too often, you will witness a pair of less-than-happy partners in the ring, perhaps the dog partner is lagging and hesitant, the human partner might be nervous and distracted.

While watching performances like this many years ago I would think, "I'll bet that they trained very hard for this event and I'm sure that they don't look like this in practice." Indeed, in my own work with my Scottish Terriers, we'd have great practice sessions, only to have things fall apart in the show ring. Watching all this, I'd ask myself, "What is the real problem here? Why does it seem so hard to transfer a great practice performance to the ring?"

In my search for the answer to these questions, I recognized that obedience competition is unique in the world of canine sports. It is the only discipline in which dog and human are performing partners and yet the dog is not allowed to be coached during the competition. I realized that handlers needed a way to close the gap between the teaching/schooling phase of their training—where they are constantly guiding, helping, and encouraging their dogs, and the testing atmosphere of the show ring—where the only outward communication permitted is a set of short commands for each exercise. My desire to help handlers and their dogs with this predicament inspired me to discover that it is possible to *build a bridge* between training sessions and the show ring.

If you and your dog are like many of my students, you will be able to build this bridge together and as you do, your communication will be enhanced, your partnership will be deepened, your performances will improve, and best of all… you'll both have more fun!

I have designed this *Bridging Process*—which is based on the *Three Modes of Training*—to allow you and your dog to develop the skills you'll need to take the personal best performance that the two of you can achieve when you have that cookie in your hand and praise in your voice and bring that same great performance into the show ring.

My co-author, Shalini, has added an important dimension to the way I teach the Bridging Process. I have always known that my dogs and I could understand each other, but like many people, I didn't fully realize how deep that understanding could be. Shalini has shown me how to trust my experience of direct communication with dogs. I find that the more I trust and acknowledge my experience, the more my intuition and understanding are validated. Working with Shalini has opened me to a deeper awareness of the connection that we all share with our dogs and has demystified this form of direct communication. By acknowledging and talking about this connection in my obedience classes, I am able to encourage others to begin to explore this amazing experience. The results of using this form of communication in our training have been happier dogs, less frustration, and better performances.

I've spent years developing a way to help people and their dogs bridge the gap between training and testing. The resulting process combines modern dog training, current education models, my professional experience with my students—both dogs and humans—and of course, my personal experience with my own dogs. Now the time feels right to spread the word, and with the help of my friend Shalini, I am very happy to be introducing the *Bridging Process* and my *Three Modes of Training* to a broader audience by offering this book to everyone who loves working with their beloved canine partners.

Shalini

I am happy to collaborate with Marsha in bringing this important information to people and their dogs. In my work as a professional interspecies communicator, I talk with many people and their dogs who desire better communication with one another. For some clients, the goal is to improve their performance in practice sessions and at obedience trials. The experiences they describe suggest that many obedience programs are not providing an adequate explanation of how to make the transition from the learning and practice stages of the work to the show ring competition. I sense the confusion and anxiety that can come from not being fully prepared for the stress of competition. In many training programs, the ability to perform with confidence during competition is rarely discussed. It is often either ignored or treated as some sort of mystical thing... as if it is something that just comes with time, or is somehow dependent upon personality or upon the human partner's innate ability to avoid nervousness at a trial.

While it's true that some are naturally more at ease during competition than others, I believe that making a smooth transition from training to showing is a *skill* that can be both taught and learned. Marsha's Bridging Process makes this transition possible for anyone willing to put in the time, and is fun for humans and dogs alike. Along the way, you'll find that the communication between you and your dog will become stronger and flow more freely.

Over the years that my dogs and I have been training with Marsha, I have come to admire the way she encourages her students to be sensitive to their dogs' experience of the training sessions. In fact, three of the most important elements of Marsha's work—the Bridging Process, Attention Training, and Progressive Classes—are based on an awareness of the dog's perspective, thereby encouraging us to strengthen our empathic connection with our dogs. Strengthening that connection results in a better show ring performance and, even more important, deepens our communication with our dogs, allowing us to share our lives more fully with one another.

The Bridging Process reminds us to take into account the dog's uncertainty and possible confusion when the handler is silent in the show ring and solves this problem by breaking up the training into three modes that the dog can learn to recognize, ultimately helping the dog gain the self-confidence necessary for a great performance.

Attention Training teaches the dog much more than simply how to pay attention to his handler. Learning to direct one's attention is a fundamental aspect of self-control. The ability to focus one's attention increases the capacity to learn new things; this is true for dogs and humans alike. And as we work through Marsha's ten stage program, teaching our dogs to direct their attention toward us, we are also focusing our attention on our dogs. Our awareness becomes more and more attuned to our dogs' focus of attention.

In the Progressive Class, dogs and handlers are not segregated by their experience or skill level. When we participate in these inclusive classes, we become aware of the wide range of encounters and experiences that can have an effect on our dogs' training. The more we take part in this type of class, the more we appreciate that the dogs in the class are thinking, feeling beings with dynamic personalities—learning from and responding to their peers and their environment as well as to their handlers.

Being aware of our dogs' experience of our work together implies that we are able to communicate with each other. It's often said that people and dogs "speak different languages." Of course, it is true that, as humans, we use spoken language and, as canines, our dogs rely more on posture and body language. But my experience tells me that we also share a common language—that is the language of thought. Part of my contribution to this book is to highlight our natural ability to communicate directly with our dogs and to show how that ability can inform our training. I believe that we all have this ability, and that to use it with awareness and to its full potential, all it takes is the desire to understand, the intention to "listen," and the willingness to put aside our doubts.

It is my hope that the ideas you find in this book will help you and your dog deepen your awareness of one another, broaden the possibilities of communication, and open to new opportunities for fun and understanding in your training and in your life together!

Building A Bridge
from Training to Testing

Introduction

THIS BOOK IS FOR people who enjoy learning and practicing the traditional obedience exercises with their dogs. The information and ideas presented here will be valuable to handlers and instructors alike. Whether you are an accomplished handler or a relative newcomer to obedience training, if you and your dog are familiar with the basic obedience exercises—such as heeling, stay, and recall—you will appreciate and be able to apply all that this book has to offer.

The important thing for you to bring along as you read this book is an eagerness to learn new skills that will give your training that extra sparkle and allow you to deepen the relationship and strengthen the communication between you and your dog.

Whatever your background, whatever your level of experience—whether it is Pre-Novice, Novice, Open, Utility, or beyond—this book will help you get more out of your training time. And, if you're an obedience instructor, you'll find innovative ideas that will enhance your lesson plans. Whether you are a serious competitor or simply train for the joy of it, by applying the tools in this book to your training program, both you and your dog will gain a deeper understanding of your work together, resulting in an improved performance in the show ring as well as an enriched partnership in your daily lives.

THIS BOOK IS a guide that will help you and your dog bridge the gap between your *practice sessions*—where you can motivate your dog with praise and encouragement—and the *obedience trial ring*—where you and your dog must work together in near silence and still deliver your best performance. The motivational practice session and the nearly silent test are two very different modes of training—especially from your dog's point of view. You can help your dog make the transition from the practice session to the test by adding an intermediate *confidence building* mode of training. When you devote a certain percentage of your training time to building your dog's confidence, you'll be helping him learn to perform without needing to rely on your constant encouragement. When you understand the distinctive characteristics of these three different modes of training and practice them with your dog, working in silence will become familiar and friendly territory.

This book will show you how to effectively employ these three modes of training. The term *training mode* refers mainly to your demeanor while you are working with your dog—you can give your dog constant, *motivational* feedback, or you can help build his *self-confidence* by giving him quieter, more subtle encouragement while letting him be more self-reliant, or you can be the nearly silent partner, giving only commands, *simulating the show ring*. While it is your behavior that distinguishes which mode you're using, your dog's response to the training mode will be just as important to the performance that you end up with as a team. Dogs are observant, intuitive beings and are almost always attuned to our body language, facial expressions, the sounds we make, and yes, even our thoughts. To be the best handlers we can be, we must communicate on all these levels with our dogs.

Of course, we can't be thinking about and doing all of these things all at the same time, we'd go crazy trying to keep track of it all. Instead, the way to "cover all the bases" is to be aware of your *intention* while training and see how it differs in each mode of training. In the first mode, it is your intention to motivate your dog by giving lots of feedback, schooling and helping him in every way. Your intention in the second mode is to help your dog build confidence in his skill and understanding of the exercise so that he begins to realize that he can give a good performance without your constant support. And in the third mode, it is your intention to become familiar with working together in near silence, with commands but no feedback, while maintaining a strong thought connection (more on this a little later). By becoming aware of your intention in each mode of training, it's more likely that you'll be communicating with your dog on all levels at once, and you'll be communicating in a way that changes appropriately as you change your mode of training. This holistic communication, supported by your intention, is at the very heart of the *Bridging Process*—the process of bridging the gap between motivational training sessions and the formal restraint of the show ring.

As you work with your dog in the three different modes of training, you'll gain a better understanding of your strengths and weaknesses as a team. You'll be able to gauge your progress with each exercise as you practice it in each of the three modes, allowing you to fine tune every exercise. The *personal best* performance to which each handler-and-dog team can aspire is summed up in this easy-to-remember formula: Dog Talent + Human Talent + Time Spent Practicing = Training Results. Having unrealistic expectations can lead to a lot of stress. We can relieve that stress by keeping our self-evaluations honest and our goals realistic.

Throughout the pages of this book, you'll notice that we occasionally alternate between the pronouns *he* and *she* whenever we're referring to a hypothetical dog, rather than using the ghastly-sounding "it" that we've all seen used in some books about animals. We hope that you'll find the many scenarios depicting dogs and their handlers working together to be informative and realistic.

We write from the perspective that our dogs are our intelligent partners—that they are sentient beings capable of complex thought, decision making, and a full range of emotions. We believe that our readers will agree that our dogs are our full partners in this endeavor and not simply extensions of our own talent or skill, nor of ourselves. When we say that the dog is a *full* partner, we are not suggesting that she is an *equal* partner. The human has the role of teacher in this partnership and as such should have an attitude of a friendly leader. When we watch dogs interact with one another, we see that a dog who is naturally an accepted and respected leader is usually one who is friendly and at ease with the other dogs and gives clear signals to those in her pack or family unit. These are traits we should always strive to emulate in our work with our dogs for they will go a long way in helping us create the true partnership that we seek.

While the human is the partner responsible for making most of the choices for the team, it is important to keep in mind that the dog partner can (and should!) take responsibility for his or her own performance. Think about it… if she's simply obeying a command, her performance might be accurate, but it will be flat. But if she really understands the exercise and has self-confidence in her knowledge and skill, if she willingly takes responsibility for her own performance, then her work will really shine. And training sessions will be more satisfying and a lot more fun—for everyone.

In fact, this is what the Bridging Process is all about—both partners really working together—and *thinking* about how they're working together. By adding that all-important Confidence Building mode to your work, you are asking your dog for more than just obedience to your command, you're asking her to use her mind, to give more of herself, to work *with* you—not *for* you. When your intention is to help build your dog's confidence in her own skills, it means that you believe in her ability and her intelligence. And so, by coming to your training sessions with the assumption that your dog has the ability to be interested and inspired, to think and be engaged, you are actually expressing your love and respect for her as an intelligent being… just think how that supports her in reaching her full potential. Indeed, just about anyone would blossom under those conditions!

We hope that readers will easily recognize that, at the heart of all the descriptions and suggestions regarding the three modes of training—at the core of all the concepts and the ideas that we offer throughout this book, is our belief that it is natural and normal to appreciate dogs as wonderfully curious beings who enjoy learning new tasks, who thrive on understanding new ideas, and who will eagerly watch and learn from us and from each other, and who may even, when given the chance, help one another during training sessions.

The alert reader will notice clues here and there that we are writing from the perspective, gained by our own experience, that a thought connection (or telepathic link) between human and dog is also a natural and normal part of life. Our experience shows us that when we become aware of the direct transference of thoughts between ourselves and our dogs, when we acknowledge our natural empathic and telepathic abilities, then we can employ that direct communication in our training—quite often resulting in a better performance. If this is a new idea to you, we strongly suggest that you give it a try. We believe that if you decide to conduct an experiment by making an honest effort to suspend any disbelief that may be lurking in your mind—that if you sincerely assume that you and your dog can easily and fluently communicate by direct transference of thought—you *will* experience it for yourself.

THIS BOOK IS NOT a training manual. While our purpose in writing this book is to help handlers and their dogs get more out of obedience training, we do not promote any particular training method or teaching style. We point this out because we'd like to emphasize the difference between *training methods* and the *three modes of training.* In using the term "training method" we are referring to the method or program that a person uses to teach their dog new exercises. Behavior shaping, classical conditioning, the many variations of operant conditioning (such as clicker training), and the model/rival technique are all examples of different methods of training or teaching a new task or behavior. In contrast, the *three modes of training* are three distinct *ways of practicing* with your dog which can be employed within the framework of the training program of your choice. However, because working in the three modes of training depends upon the handler and dog thoughtfully responding to one another as a team, only those training methods that are based on mutual respect and trust will be in harmony with this approach to training. If you've read this far, then you should be able to correctly guess that any training method that resembles the old-fashioned, dominance or force-based training methods would not be suitable for use with the three modes of training. (Thankfully, we can say that those compulsory, fear-driven training programs are fading fast as the many more enlightened teaching methods gain in popularity.)

It is important to point out that this book does not present scientifically evaluated evidence of the effectiveness of any of the techniques or concepts that are presented. We are not research scientists. Our intention is simply to share with you, the reader, many of the ideas and practices that have grown out of the experiences we've had over the years with our own dogs and with our clients and their dogs. In addition to the main focus of the book—the three modes of training—this book also touches on topics such as social learning, attention training, and telepathic communication, as well as other suggestions and advice that we believe can help create an environment conducive to bringing out the best in dogs and their handlers. Everything we offer the reader is based on our own experience—we have seen these ideas and practices contribute to countless successful partnerships both in and out of the obedience ring.

We'd like to focus for a moment on a phrase that we use throughout the book—*the Bridging Process*. We realize that the use of this term could initially cause some confusion to readers who are familiar with the science of animal training. It is probably already clear from our discussion so far, but for clarity's sake we should point out, that when we refer to the three modes of training as the *Bridging Process*, we do not intend any connection to the operant conditioning terms, *bridge* or *bridging stimulus*. Within the field of operant conditioning, these terms refer to a very specific type of feedback. A *bridging stimulus* is a positive reinforcement (such as a sound or word) which lets the animal know that he or she has performed a task correctly. These terms are in no way related to our use of the expression *Bridging Process* to refer to working in the three modes of training—Motivational, Confidence Building, and Testing. Over the years, as Marsha was developing the three modes of training, it became clear that the process that was emerging from her work was *bridging the gap* between training and testing—for her clients and their dogs as well as in her performances with her own dogs—and this distinctive approach to practicing the obedience skills became known to everyone who worked with Marsha as the *Bridging Process*.

THE UNDERLYING MESSAGE OF THIS BOOK IS one of creating partnership with our dogs based on love, friendship, and mutual respect. Humans and dogs can easily and happily live together, in part, because it is natural for both primates and canids to live within social or family groups in which each member has a role to play. Participating in obedience classes and trials gives the intelligent dog an interesting and challenging job. It helps him feel needed and wanted and that he is capable of making a contribution to his family. An obedience-trained dog is an ambassador for his kind—and sometimes even for obedience training itself. Well-behaved dogs are welcome guests in many places and help make life better for all dogs by demonstrating that they can be good citizens. Meeting a well-trained dog is often what persuades a person to enroll in a class with their own dog for the first time. Our lives are enriched by living with dogs who have developed their intelligence and their skills—and it's fun to live with a dog who will happily retrieve a dropped glove, collect the dog toys from the yard, or help by carrying a bag. Just as important, our dogs are safer when they will drop an unknown object from their mouths or stop and stand on command.

In so many ways, learning and practicing the obedience exercises with our dogs, whether we ever participate in competition or not, helps us develop a deeper understanding of one another—and is also a fun way to spend quality time together. We're confident that the readers of this book will share our point of view on the relationship between training and showing—we show because we train, not the other way around. If tomorrow morning we woke up and discovered that there would never be another obedience trial, it wouldn't change our love of training or our desire to work with our dogs. In other words—it's far more fulfilling to *show to train* than to *train to show*.

Of course it's true that some don't compete because of anxiety. In that case, working in the three modes of training can often transform such a reluctant dog or handler into an enthusiastic competitor. That said, competition just isn't for everyone. Some people love to compete and others have no interest in it whatsoever. Some dogs enjoy the excitement of showing and others do not. It's worth noting that, as the authors of this book, the two of us represent the opposite ends of the spectrum. Marsha enjoys the challenge of competition while Shalini would rather have a quiet weekend at home—but both of us love to work with our dogs.

Obedience training improves our lives with our dogs in so many ways. We're sure that if you love working with your dog, you'll find satisfaction in working in the three modes of training, whether or not you are preparing for competition. The Bridging Process helps make training more interesting for your dog and more rewarding for you.

Chapter 1

The Bridge

Three Modes of Training

The bridge from training to testing consists of three distinct modes of working with your dog. These modes are three different ways of communicating with your dog; each mode representing a unique approach to performing the individual obedience exercises. When you put them all together, you will have bridged the gap between practicing in your own backyard—where you are free to help and motivate your dog—and the show ring—where you are not permitted to help or encourage your dog in any way.

We believe that, like us, most people participate in obedience training in order to spend quality time with their dogs. Obedience titles are a great acknowledgment of all our hard work together, but for most of us they are secondary in importance to having fun with our dogs and finding enjoyable ways of sharing our lives with them. By consistently practicing the obedience exercises in the three modes of training, handlers and dogs develop better communication with one another. This not only improves our scores, it also enriches our everyday lives, helping us make the most of the precious time we have with our beloved canine partners and companions.

The three modes of training that make up our bridge are distinct styles of practicing that you and your dog will come to recognize, they are: the **Wow** (Motivational) Mode, the **Confidence Building** Mode, and the **Testing** Mode. As we learn and practice the obedience exercises in partnership with our dogs, we use these three modes to motivate, to gain confidence, and to test our performance of each exercise.

Wow—motivation, schooling, helping, rewarding

The motivational mode of training is called the *Wow* mode to remind us that training sessions should be fun for both dogs and handlers—with happy dogs thinking, *"Wow! This is great!"* and handlers enjoying and appreciating their dogs' performances, *"Wow! What a good dog!"* Because motivation is so important when learning something new, we always work in the Wow mode when teaching or learning. But Wow is not just for when we're learning new things. This motivational approach to training is the mainstay of all our work with our dogs. By far, the greatest percentage of our training time is spent in the Wow mode. In this mode of training, we are continually schooling our dogs, giving lots of feedback with voice, treats, and body language. The support and encouragement of Wow keeps the performance of the exercises fresh and exciting for our dogs and for us. But, as we all know, we cannot give our dogs any outward feedback during a performance at an obedience trial, so we need to connect the inspiration and support of Wow to the silence and self-reliance of the show ring. The *bridge* that makes this connection is the Confidence Building mode of training. When our dog has gained a good understanding of an exercise and can *easily* perform that exercise in Wow, we can begin to bridge that exercise. The Wow mode of training should always make up at least 70% of your training time.

Confidence Building—limited compliments and corrections

As the name implies, our intention in Confidence Building is to assist the dog in building his confidence in his performance of the exercise, so that he understands that he can do it without the constant feedback from his human partner. In the Confidence Building mode, we use quiet compliments and gentle corrections. In contrast to the steady stream of feedback and the extravagant encouragement of Wow, the more subtle Confidence Building mode is characterized by a calm feeling of teamwork, with occasional, discreet assistance rather than continual schooling. The handler is focused on creating an atmosphere where the dog can build confidence in his ability to be a full partner in the performance of the exercise. The concentration required of both dog and human in Confidence Building takes a lot of energy, so we don't spend very much of our training time in this mode, we always return to the support of the

Wow mode for the majority of our time together. Confidence Building will represent 20% to 30% of training time.

Testing—silence, planned jackpots

Finally, when our dog is confident in her expertise with a particular exercise, reliably performing it well in both Wow and Confidence Building modes, we can plan a Testing session for that exercise. In Testing, we simulate the obedience trial ring. By working in silence, as we do in the show ring, both dog and human become acquainted with the minimal interaction that makes the show ring so disconcerting to many obedience trial participants. But the all-important difference between the Testing mode of training and the actual show ring is the addition of the planned jackpot—a short break in the action during the test. By adding a planned jackpot to the silence of the Testing mode, we help relieve the tension of that silence and also create a sense of expectation that helps the dog keep her focus—because when a dog who is experienced with the three modes of training recognizes the silence of the Testing mode, she knows that there may be a cookie coming soon and keeps her attention on her handler. Used very sparingly, the Testing mode of training will never represent more than 10% of your total training time.

The thing to remember is that—by far—most of our training time is spent in the Wow mode. We want to keep our work together happy and fun… after all, isn't that why we're doing all this in the first place? Wow is where we have fun with our dogs and keep our performance happy and fresh. As important as the Confidence Building and Testing modes are in bridging the gap from training to the show ring, these modes require a focus and mental energy that, if overdone, would sap the strength of both human and canine partners. Therefore the Confidence Building and Testing modes must always represent the smaller percentage of our work with our dogs. A typical week of training sessions for an experienced handler and her dog, who have polished their performance and are preparing for an obedience trial, might be 70% Wow, 20% Confidence Building, and 10% Testing. We'll talk about these percentages in more detail a little later.

Wow–Confidence–Testing vs. Beginner–Intermediate–Advanced

Most of us are accustomed to thinking in terms of the three stages of learning and perfecting a skill as **beginner, intermediate, advanced**, and we expect students to be ranked by their level of skill and then put into a class with others of the same level. Obviously, we all—dogs and humans alike—go through those three stages as we learn new techniques and skills. However, there is a difference between teaching a new technique and practicing it once it's already understood.

As we learn and perfect our skills, we are also continually practicing what we've already learned. So instead of thinking of our work with our dogs as just a learning process (where we would simply move through the beginner, intermediate, and advanced levels), the three modes of training help us see our training as both *a learning* and *a practicing* process.

These three modes of training are quite different from the familiar beginner, intermediate, and advanced levels of learning. As we have seen, the Wow, Confidence Building, and Testing modes of training are three distinct ways of practicing each exercise with your dog. By dividing the time you spend working with your dog into the three modes of training, you'll be doing much more than simply learning and repeating exercises. You and your dog will be intentionally developing confidence in and polishing your performance of the obedience exercises as well as strengthening your relationship and enjoying your partnership.

Important Point!

The Three Modes of Training: **Wow–Confidence–Testing**

are **not** the same as

The Three Levels of Learning: **Beginner–Intermediate–Advanced**

Chapter 2

Defining the Three Modes of Training

The three modes of training are three distinct ways of working with your dog, and we've looked at how they differ from levels of skill, such as beginner, intermediate, and advanced. In this chapter, we further define Wow, Confidence Building, and Testing and look at the unique characteristics that distinguish each one. And later on, in Chapter 7, we'll explore the three modes of training in action by looking at detailed examples of the recall and heeling exercises performed in each of the three modes of training.

Wow - The Motivational Mode

In building our bridge from training to testing, the greatest percentage of training is done in the motivational mode of training, or "Wow" as we call it. No matter how advanced, no matter how well you and your dog know the exercises, motivation is the key to keeping everything fresh and exciting, whether you are practicing your best exercise or just beginning to learn something new. With lots of vocal reinforcement and plenty of treats your dog will be thinking, *"Wow! This is fun!"*

In the Wow training mode, you give your dog lots and lots of feedback so that she always knows how she's doing. A dog, like anyone, doesn't like to have to guess about how she's doing—and in Wow there's no guessing. In this mode, your constant encouragement motivates your dog, and your consistent schooling and moment by moment feedback keeps her from having to wonder how she's doing. In Wow, she should be able to easily tell if she has done it right, or done it wrong—dogs really want to know either way. If she's doing it right, you're quick to encourage and praise. And if she starts to get it wrong—for instance, she's headed for an obviously crooked sit—you are ready to gently interrupt her with a soft nudge in the correct direction and a voice prompt, *"Straight,"* followed by a happy, *"Yes, that's better!"* when she ends up in a straight sit.

One of the most heart-breaking things to see in a practice session is a dog who is trying so hard, and is doing a great job, but is looking up at her silent handler with that confused expression, *"Is this what you want? Am I doing it right?"* When you train in Wow, with plenty of feedback, your dog will always know how she's doing, and that makes for a happy dog!

When you're working in Wow, you are assisting your dog on many levels; you are schooling her, you are helping her understand what you want by giving obvious visual cues, you are rewarding her performance with happy praise, and you are helping hold her attention with a variety of cues including food, voice, visible markers, body language, and toys or other props.

Your voice plays a powerful role in the Wow mode, allowing you to communicate your approval of her performance and your admiration for her effort. As your dog's partner, it's natural for you to be emotionally involved in the work. It's natural that you should be *thrilled* when you realize that she has understood something new—when you see that "light go on." And although you're always communicating on many levels, it's largely through your voice that she will feel the energy of your excitement and appreciation. Think of your dog's performance as her gift to you, and when your dog has just done a great job, it is your sincere praise and encouraging, heartfelt phrases such as, *"That's great!" "You're so smart!" "I know you understand this!" "What a clever girl!"* that can be your way of giving her a gift in return. The sincerity of your praise is important in all your training, and in Wow you should take every possible opportunity to use it. On the other hand, dull or robotic praise tells your dog that you're not engaged in the work and, worse still, praise that has a sarcastic tone can actually hurt your relationship and be destructive to your training. By letting your joy show, by showering your dog with *sincere, heartfelt* praise when she does a good job, you'll be reinforcing her understanding of the exercise and strengthening your partnership.

Often, in the Wow mode, we will break up an exercise into segments in order to make it easier to school. Examples of this are things like separating the front from the finish, walking in to give a treat during an exercise—such as after a "go-out" during the directed jump exercise, or practicing heeling in short segments.

Wow is all about motivating your dog and generating enthusiasm for your work together and that means keeping the lines of communication open and flowing with constant and consistent feedback! So if there's one thing you should remember about the Wow mode of training—it's that your dog should never be wondering whether or not you're happy with her performance.

The Confidence Building Mode

OK, so at this point you might be thinking, "Supporting my dog with motivation and feedback sounds great, but how do I get from there to the show ring where my dog can't rely on me for anything like that?" The Confidence Building mode bridges the gap between your motivational support during practice and your dog's self-reliance in the show ring or in the Testing mode of practicing.

Once you and your dog are performing a particular exercise *without struggle* while in Wow—meaning that you can rely on him to have good form, without hesitation or refusal—then you are ready for Confidence Building, but just in that one exercise. As the name implies, the goal here is to build your dog's confidence in his own ability to perform the exercise. This is where you build that bridge to a self-reliant performance.

In the Confidence Building mode, your feedback becomes very subtle, but—and this is an important point—your feedback does not disappear altogether as it does in the show ring. Your voice should be low-key and discreet. The treats are hidden and given out much more sparingly, and there are occasional soft compliments, but only given *after* the dog completes a task, with a quiet, *"good…,"* or a whispered, *"that's right."*

Not only is your feedback to your dog more subtle in Confidence Building than in Wow, it is also far less frequent. There's not a running commentary coming from you about how your dog is performing as there is in Wow. In the Confidence Building mode you're just occasionally giving your dog quiet compliments and gentle corrections. By silently letting the dog perform an exercise or a segment of an exercise on his own, he gains the self-confidence that he'll need in competition. In Confidence Building, we don't take the feedback away completely, but the compliment or the correction comes out of the silence after the dog has had a chance to perform on his own. So, instead of encouraging and helping him with a happy *"Yes!"* during the body of the exercise, you would silently wait until he has performed and then offer a quiet compliment, *"Good boy!"* Likewise, instead of interceding as he is in the process of making an error, as you would in Wow, in Confidence, you wait silently as he sits and then immediately reach down, gently nudging him over while softly saying, *"Straight. That's better!"* (In Chapter 4, we will describe corrections in both Wow and Confidence Building modes in much more detail.)

All this is done with the intention of helping your dog understand that, while most of the time you'll provide him with plenty of feedback (remember, you'll be in Wow more often than any other mode), there are going to be times when he'll be expected to rely more on his own knowledge of the exercise than on your motivation. The Confidence Building mode helps him begin to get used to this idea.

The Testing Mode

In the Testing mode, just like in the show ring, there is no verbal communication other than the specific commands for the exercise and there is no praise given during the exercise. It is this lack of outward communication that usually brings the dog down and perhaps even makes her nervous. She might be thinking, *"Why did Mom stop talking to me? Did I do something wrong?"* Those can be very distracting thoughts for your dog and usually, the more she thinks like that, the worse her performance gets.

The cure for this distraction is the planned jackpot, or what we call the "Lotsa"—as in *lotsa* cookies, *lotsa* fun, *lotsa* love! Everyone wants *lotsa* this good stuff in life, right? Well, that includes your dog!

The key here is that the jackpot is planned for ahead of time, it is *not a reward* for getting it right. A reward is a type of feedback and the very thing that makes the Testing mode distinct from the other modes is the lack of feedback. The jackpot is a tension reliever, and as such it needs to come out-of-the-blue from your dog's point of view.

For example, before beginning a heel pattern in the Testing mode, you will decide where in the pattern you will break out and do a jackpot—when you get to that point in the pattern, you suddenly stop and shout, *"Lotsa cookies!"* or whatever happy phrase you'd like to use to signal a break-out to your dog. Spend a bit of time reaching for the hidden treats and make a really big deal about it; this will relieve the tension that may have been building during the otherwise quiet heel pattern. After she eats the treats, have her get back into heel position and continue with the pattern *as if the break-out hadn't even happened*. Again, the important thing to remember about the jackpot is that it is planned for ahead of time, it is never a reward for your dog's performance.

In addition to the planned jackpot, you'll plan ahead for the performance level that you will accept from your dog. If your dog makes a mistake during the test that would result in a non-qualifying score during a competition, then you'll want to instantly abort the test and go back into Confidence Building mode where you can offer a timely, helpful correction and bring her back up with praise when she gets it right.

Planning ahead for your accepted performance level requires an honest evaluation of you and your dog's capability as a team. As Marsha frequently reminds her students, the results of your training are based on:

Dog Talent + Human Talent + Time Spent Practicing,

where *talent* is the capacity or ability to perform the exercises, a combination of natural aptitude and learned skills. When you make a decision regarding the performance level that you'll accept in the test, be realistic and keep in mind the individual skills and talents of both you and your dog and consider the amount of time that you spend practicing on a regular basis. These are the major factors that will determine your performance during the Testing mode—and in actual competition.

If you're not all that serious about competition, and you'd be thrilled with any qualifying score, then you would only abort your test in the case of an error that would result in a non-qualifying score and nothing less. If, on the other hand, you and your dog are the high-achieving types and you are entering a competition next week aiming for a High In Trial, then you'll be ready to abort the test for even the smallest slip-up. Most of us will fall somewhere in between these two extremes. By the time we are ready to begin Testing some of our exercises, we will certainly have come to know our dog's (and our own) strengths and weaknesses well enough to make this judgment.

So, as you can see, a big part of the Testing mode is in the planning—thinking about the optimum performance level and planning where to breakout for your jackpot.

In the Testing mode of training, as well as in the show ring, you *can* help your dog by *thinking* the same encouraging words and phrases you would be using if you were working in Wow. By thinking these familiar phrases, you'll get into a "Wow-state-of-mind" even though you're outwardly in the silent Testing mode—and you'll be creating a supportive atmosphere that will help your dog stay relaxed yet motivated. (Note: If you don't already do this naturally, you may be surprised how much *thinking to your dog* helps during a test!)

Once you start doing this quiet Testing work with plenty of planned jackpots, it doesn't take long for your dog to recognize your silence as a good thing. She'll come to realize that when she hears only your commands and no feedback at all from you—it means that, at any moment, you might break the silence with a happy shout, *"Lotsa cookies!"*

You'll be surprised how fast your dog will learn to recognize the Testing mode and, instead of feeling apprehension when she notices your silence, she will be excited by it, anticipating those jackpots at any moment.

All Three Modes

It can be helpful to compare the three modes of training by looking at them in terms of the intention of the handler. In Wow, your intention is to help and encourage your dog—using constant, sincere praise and schooling and lots of feedback—assuring that your dog always knows how she's doing. In the Confidence Building mode, your intention is to allow your dog to work on her own—enough to build her confidence in her own ability—using only occasional, quiet compliments and gentle corrections. In Testing mode, knowing that your dog understands her job fully, your intention is to be your dog's silent partner, maintaining a strong thought connection between the two of you while planning ahead for the jackpot and your targeted performance level.

> *Here's a good way of visualizing the relationship between the three modes of training:*
>
> In your work together, you and your dog are building a wall of training bricks for each exercise. Every time you practice an exercise in the Wow mode, the motivation and feedback you are providing is like adding two bricks to the wall. Each time you practice an exercise in the Confidence Building mode, your dog has to rely on her own confidence a bit and so it is like adding only one brick to the wall. But each time you perform an exercise in the Testing mode, you "use up" some of that motivation and confidence and it is like taking a brick off of the wall. So, in order for the wall to be solid and strong and dependable, you'll always do at least 70% of your practice exercises in Wow, a few in the Confidence mode of training, and just a very little bit in Testing.

As you and your dog become comfortable with the three distinct modes of training, you will see how the Motivation–Confidence–Testing model will help the two of you achieve a happy, accurate performance during practice and be able to easily transfer it to the show ring. And in doing so, you will not only achieve a more dependable performance, but by working in partnership, you and your dog will gain better communication and deepen your understanding of each other. And after all, isn't that what it's all about?

Chapter 3

Pie Charts & Percentages

Moving Through the Different Modes

Now that we have defined the three different modes of training, we need to look at how to use them. We've already mentioned that there are ideal percentages—but just how do you decide when to work in a particular mode?

Always Teach in Wow and Then Keep Coming Back!

When we're teaching our dogs a new exercise or the new stage of an exercise, we always work in Wow. It's important to support your dog with as much motivation as possible during the learning phase of training, whether it's with a puppy who's just getting started or with an older, more experienced dog who's learning something new.

So, for a young or inexperienced dog just starting out, the Wow mode will be 100% of his training time. And for the more knowledgeable dog who's just learning a new exercise, *the new exercise* will be done in 100% Wow—until he masters it.

As we've mentioned before, Wow is also the mode that we work in most of the time—never less than 70% of training time. Even after the dog has perfected the exercise, he still appreciates that motivational attitude in his handler, keeping his spirits high and making the work fun and exciting.

No Struggle = Go To Confidence Building

When you and your dog can reliably perform an exercise smoothly and with no struggle while working in the Wow mode, then it's time to try it out in Confidence Building mode. But don't overdo it; taking away that motivational feedback needs to be done slowly and carefully. Remember—the intention is to gradually build your dog's confidence in his own ability, not to quiz him on his knowledge. In this mode, you are gradually helping your dog build his self-confidence by allowing him to perform the exercise on his own, occasionally stepping in with gentle corrections and quiet praise. This subtle feedback is only offered intermittently, so that he learns to trust his own judgment.

Perhaps you perform a particular exercise with your dog ten times over several days of training sessions. Only one out of those ten times will be in Confidence Building in the beginning, then you can gradually bring that percentage up a bit. But the highest percentage of time spent in Confidence Building should never be more than 30%—and that high point of 30% in Confidence represents the training time of a handler and her dog who have perfected their performances in the Wow mode for all of the exercises they know.

In fact, 30% Confidence may even be a bit too high for some dogs. There are dogs who will always require a bit more of the motivation they get from the Wow mode of training—so be sensitive to your own dog's needs.

Ready to Test?

When those subtle corrections in the Confidence Building mode are no longer needed for a particular exercise, when you know that your dog completely understands the exercise and will perform it beautifully in Confidence almost every time, then you're ready to test that exercise. At this point, you'll add a test for that exercise out of your 30% slice of Confidence, while still keeping your percentage of Wow at 70%.

In other words, if you perform this exercise twenty times over several days—you'll be Testing it only once, practicing it in Confidence Building for five of those twenty times, and returning to Wow for all of the other fourteen times. And during that test, always keeping in mind that you're ready to abort the test, if necessary, and go back into Confidence in order to make a gentle correction and bring your dog back up with that all-important praise.

The need to abort a test should be a very rare thing. If you find yourself aborting tests with any frequency, then you are either trying to test too often—in which case you've just learned that your dog needs a greater percentage of the Wow and Confidence modes—or you are Testing too soon—before your dog has the knowledge and self-confidence to perform the exercise on his own, in which case he may need some remedial work on that exercise.

Finally, when you and your dog have perfected all of the exercises that you know, the Testing mode should still not make up any more than 10% of your training time—with another 20% of your time spent in Confidence Building mode and that all important 70% in the motivational Wow mode.

The Pie Charts

The pie chart is a great way to illustrate the ideal percentages of the three modes of training. And while these percentages are not meant to be an exact science, as you come to think of your training in terms of these generalized pie charts, it will help you to naturally cycle through all three modes and not get stuck in any one way of training. Our work with our dogs should always be changing—constantly "circling" around the pie chart—from the motivational mode of Wow to Confidence Building to Testing and back again. By keeping the pie charts in mind, you'll always be aware that it is essential that you work in Wow at least 70% of your training time—the importance of all that motivation cannot be overstated!

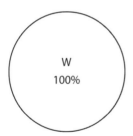

100% Wow
Training sessions for puppies or dogs who are new to obedience training are 100% motivational. You'll also use 100% Wow while teaching a new exercise.

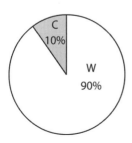

90% Wow / 10% Confidence
This is an interim stage... when just a few of the exercises that you and your dog are working on become more finely tuned and can be performed reliably and without struggle, those exercises can begin to be done in the Confidence Building mode—so we add a little slice of Confidence to our pie chart.

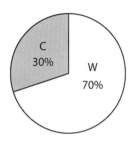

70% Wow / 30% Confidence Building
When all of the exercises that you and your dog know can be performed without struggle, the percentage of training in Confidence Building mode grows to about 30%. And from now on—for this set of exercises—the percentage of training time spent in Wow will remain at 70%.

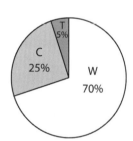

70% Wow / 25% Confidence / 5% Testing
Another interim stage… when a few of your best exercises can be performed in Confidence Building with very few (or no) corrections, it is time to begin Testing those exercises—so a little slice of the Testing mode appears in our pie chart. Notice that the time spent in Testing comes out of the percentage previously performed in Confidence so that you can maintain your training in Wow mode at 70%.

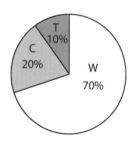

70% Wow / 20% Confidence / 10% Testing
This is a typical pie chart for an average handler and dog who have polished all of the exercises that they know and are able to test every exercise that they can perform together. When you get to this stage, you're ready to begin preparing for competition.

Chapter 4

Training Advice & Suggestions

In this chapter, we offer a few ideas that we hope you'll find beneficial in your training. We begin by sharing our thoughts on a subject that has become something of a contentious issue in the training world in recent years—communicating disapproval to our dogs. Next, we discuss the use of food in training. Most handlers use food in their training, but do they think about how many different messages those treats might be sending to their dogs? We point out that how we use the food matters and describe the distinctions between some of the different uses. We also highlight the importance of being alert for signs that a dog is feeling confused or overwhelmed during a training session—and the necessity of being quick to back-step to an easier level of the exercise or changing your mode of training in order to remedy the situation. We complete our assortment of training suggestions with some advice on keeping a record of your progress, along with a sample page from a training log.

The "C" Word—Correction vs. Punishment

Inevitably, there will be times when we are less than happy with our dog's performance. When those situations arise, we need to be able to express our constructive criticism with clarity and support, while assuring that the foundation of our partnership, our trust in each other, remains alive and well. In this section, we describe our ideas regarding such supportive corrections in the context of the three modes of training. Please keep in mind that the type of correction we are describing here is meant only as *feedback for the dog who already knows and understands the exercise.* We will not be addressing the use of corrections in *teaching your dog something new.* It is not our intention to discuss or compare methods for teaching new behaviors. There are many successful methods of teaching new skills; some include negative reinforcement, although many do not. It's up to you, as your dog's primary teacher, to find the teaching method (or combination of methods) that works best for the unique team of you and your dog.

Before we go any further, we'd like to explain what we *don't* mean when we use the word *correction*. This word seems to mean different things to different people. If the term *correction* is used synonymously with words like *punishment* and *compulsion*, then it's no wonder that it has become something of a dirty word as the world of dog training has shifted toward positive methods. From the dog's point of view, both punishment (the inflicting of pain or some other penalty in retaliation for bad behavior) and compulsion (the use of physical force) create stress and fear which lead to a desire to avoid the entire situation. Obviously, this is not the meaning that we have in mind!

The first thing to realize about corrections in the context of the three modes of training is that they will only occur in either the Wow or the Confidence Building mode. A correction cannot happen while you are working in the Testing mode since the Testing mode is defined by the complete lack of feedback. So, if you did choose to give a correction during a test, you'd be aborting the test and going back into Confidence in order to give the correction.

Therefore, within the framework we've just described, our general definition of a correction is *the interruption (using voice and touch) of an incorrect behavior with the intention of giving the dog feedback about his performance, followed by praise for getting it right.* This general description fits for both Wow and Confidence Building modes, with the most significant difference between a correction in Wow and a correction in Confidence being the *timing* of the interruption.

The *intention* of the handler is an important part of the definition, as is the *praise*. When giving a correction, our intention is to interrupt the error and yet still have the dog's attention remain on the exercise and the handler. In other words, our interruption of the behavior, as well as the subsequent praise, should be *mild enough* that it won't disengage the dog from the task. The dog should understand that the correction is meant to be helpful feedback about his performance. The interruption itself is done with both voice and touch.

Here's an example:

> You are working in Confidence Building mode. It is the midpoint of the drop-on-recall exercise during an obedience class and the instructor has just given you the signal to drop your dog. You give your dog the cue to drop, and instead of dropping as usual, he just stops and stands there, looking at you as if to say, *"Isn't this good enough? I stopped, do I really have to lie down too?"* In answer to his question, you walk over to him, repeat the cue, *"Down,"* while touching him with stiffened fingers between his shoulders with a downward pressure. When he drops into the down position, you compliment him softly, *"Yes, that's a better down."* Then repeat the exercise in Wow with lots of feedback—only going back to Confidence when his drop is again consistent in Wow.

The strength of your voice and touch depends mainly upon the sensitivity of your dog, using *just enough* firmness so that he understands it as a correction, but not nearly so much that he gets distracted from the task at hand, or even worse, that he wants to avoid you or the exercise all together.

The *touch* is important for the emphasis that it adds. Think about how much more meaning a simple touch can bring by picturing this scene:

> You are walking down the street with a friend and she says, *"Watch out for that bump in the sidewalk,"* but just at that moment, something across the street catches your attention, you miss the meaning of her warning and you trip and nearly fall.

But what if she had touched your arm as she spoke? It's nearly certain that your attention would not have wandered, you would have heeded her warning and easily avoided the hazard—that little touch makes all the difference.

Again, a correction is supportive feedback for the dog who completely understands the exercise and whose error is simply the result of a sloppy performance or of being distracted. We only correct a dog who knows and understands his responsibility to such an extent that a mild interruption, combining voice and touch, will be enough to prompt him to realize and correct his mistake. A sophisticated, well-educated dog who enjoys getting it right will want to know when he didn't. When you give a good correction, your dog should be thinking, *"Oops, I didn't get that right...,"* and then, as he offers a better performance he'll be thinking, *"...is this what you want?"* And this is why the correction isn't complete until you praise and reward him for the corrected performance. Because if you haven't answered his question, *"...is this what you want?"* you will, at the very least, have undone some of your training, and at the worst, have undermined some of his trust. Think about how you'd feel if, at work one day, you made a mistake that obviously disappointed your boss (whom you like and respect, of course). But you realized your error and then put in a huge effort to correct it, outdoing even your usual high standards. You're anticipating a compliment or some sign of appreciation or acknowledgment, but your boss says nothing. Pretty yucky feeling, isn't it? And so it is for the dog when his handler forgets to *finish the correction* with praise. This point cannot be over-emphasized. Always, always, always... finish your correction with praise!

Timing is Everything—
Comparing Corrections in Wow and in Confidence Building

So far, we've described what we mean (and what we don't mean) when we use the word correction. Now let's look at how a correction in the Wow mode of training differs from a correction in Confidence Building. When you're working in Wow, you are giving your dog constant feedback about her performance, so that your dog will never wonder whether or not you're happy with her performance. And in Confidence Building mode, you use occasional quiet compliments and gentle corrections in order to support your dog while she is building confidence in her own skills and ability to perform. The most significant difference between a correction in Wow and a correction in Confidence is in the timing.

Remember that we use the motivational Wow mode of training both when we're *teaching* our dogs something new and also when we're *practicing* exercises that they already know. As we mentioned above, this is not a discussion of teaching methods. Our aim here is to describe the type of correction that we would use in Wow, with the intention of helping our dogs do a great job with exercises that they already understand. The timing of a correction in Wow is important. In keeping with the constant feedback of Wow, the handler interrupts the dog's mistake *as soon as* she sees it beginning to happen. In the case of a crooked sit during a heeling exercise, the handler would intercede as soon as the dog is obviously swinging outward into the wrong position, but is not yet sitting down. While the dog is still in motion, the handler reaches down with a gentle touch saying, *"Straight."* As the dog corrects the sit by responding to the touch and voice, moving over into a straight sit, the handler completes the correction with, *"Good straight, yes!"*

In the Confidence Building mode of training, we are helping our dog gain confidence in her own knowledge and skills. In order to do that, the handler is quiet most of the time, allowing the dog to work on her own for the most part so that she knows she can do it without continual feedback. In Confidence Building, corrections (like compliments) are given only occasionally, and come later in relation to the dog's performance than in Wow. While the correction is still an interruption of the flow, it is given just *after* the dog's mistake—rather than at the moment you see it start to happen, as you would in Wow. The handler silently lets the dog complete the exercise or segment of the exercise and then immediately either offers a quiet compliment or a gentle correction. Perhaps the dog did an especially perfect sit at the halt in a heeling pattern—her handler could softly say, *"Yes!"* or *"Beautiful!"* as soon as the dog is actually sitting. If instead the dog's hind end had swung out from heel position, resulting in a crooked sit, her handler would wait until the dog was sitting and then give a gentle correction consisting of a nudge to the dog's left side and soft voice, *"Straight,"* followed by, *"Yes, that's better,"* complimenting her on getting it right by moving back into heel position. The important thing to remember about a correction in Confidence Building mode is that it comes out of the silence *after the dog's error*. You offer the correction by interrupting the flow of the exercise *after* she hesitates or gets something wrong, stepping in to gently correct her—with a verbal cue and a light touch—always remembering to finish with that all important compliment!

Do Not Repeat

Whether in Wow or Confidence Building mode, a correction should be a rare occurrence. A correction that is used repeatedly ceases to be a correction and becomes nagging. Nagging doesn't help anyone. If multiple corrections *seem* to be needed, either the dog is being overwhelmed with a task that she doesn't understand well enough—in which case she needs more schooling—or, the handler is spending too much training time in Confidence Building mode and should include more motivational training. In either case, the answer is to go back to Wow mode for either more schooling or more motivation.

A Reminder

The urge to punish or to use compulsion usually arises from feelings of anger and frustration. As we all know, either one of these feelings is a red flag. We're sure that readers of this book know that a basic guideline in working with dogs is to never, ever train when you're feeling angry or frustrated. We all feel these things at times; it's natural. Experienced handlers know that feelings of anger or frustration, and the actions that arise from them, will wreck your training efforts every time. But it never hurts to remind ourselves that whenever these dreadful feelings intrude into our training, it's time to end the session immediately and begin again later when we're feeling happy and relaxed again.

Uses for Food in Training

There are several different ways to use food while working with our dogs. In order to use food most effectively, we need to be able to recognize the distinction between the various uses from the dog's point of view. The differences that dogs tend to notice depend on things like the timing of receiving food in relation to the performance of a task, whether we move the food or hold it stationary, whether the food is visible or hidden, even the amount or type of food given. In this section, we describe a few categories of uses for food and the characteristics that distinguish them.

The food that you use in training should be something that your dog really loves, a very special treat such as fresh meat or cheese. Avoid using crunchy treats for uses such as working rewards, lures, or markers. Food used in these ways is given during actual training and if the dog has to stop what she's doing to chew or crunch the treat, you could lose the continuity of the exercise.

Food as a Working Reward

A **working reward** is given *immediately* following a good performance of an exercise or a segment of an exercise. Or, when teaching something new, it's given just as the dog has offered the early stages of the correct behavior. When food is used as a reward for a job well done, it is actually secondary to the reward your dog receives almost instantly by *hearing* the pleasure in your voice, *"Yes!"* and *feeling* your sincere appreciation for her effort. A food reward reinforces the appreciation your dog needs to feel coming from you so that she knows for sure that she got it right. Timing is crucial in a **working reward**. And the timing of the reward is not just in the immediacy of giving the food, it also refers to the amount of time it takes the dog to eat it. A **working reward** should be soft and small, able to be swallowed quickly so as not to distract from the exercise, for example—a piece of meat or cheese about the size of your pinky fingernail. If the piece of food is big or crunchy, your dog will have to stop to chew—and in the extra time it takes to stand there crunching a big cookie, you'll end up with a big break in the action, effectively disconnecting you both from the task at hand.

Food as a Release Reward
There are times when you want to release your dog as a reward. This can come at the end of a work session, after a particularly long or difficult exercise, or whenever you want to stop working and celebrate an especially great performance. The timing of the food in a **release reward** is a bit slower than in the **working reward**. Again, the food is secondary to your voice, which in this case is announcing a break in the action, *"Yay! Great job! Good girl!"* The food you give during a release can be larger in size and greater in amount than in the **working reward**. You could even take a moment or two to dig into a pocket to get it out while you're talking it up, *"Good girl! That was great! Do you want some cookies?"* this creates anticipation which adds to the fun. A **release reward** can also include toys and games, but here we're just describing the use of food for this purpose.

Food as a Jackpot

Here, we are using the term "**jackpot**" specifically as it applies to the three modes of training. A **jackpot** is the food that's given at a planned break-out point during your dog's performance in Testing mode. Several things make a **jackpot** different from a reward. A **jackpot** is intended to be a break in the action, so unlike a **working reward** which is a tiny treat slipped to the dog quickly, you can take your time digging it out of your pocket. Another difference between the **working reward** and the **jackpot** is that, rather than one quick tiny mouthful as in a reward, the **jackpot** can consist of several small treats. So the **jackpot** is actually somewhat like the **release reward** in that it can be several small treats and the timing can be relatively long. But the most important characteristic that distinguishes the **jackpot** from either type of food reward is that its timing is planned for ahead of time and is not in any way a response to the dog's performance. Therefore, by definition, the **jackpot** is not a reward.

Visible, Moving Food as a Lure

When you move your hand while holding food that is visible to your dog, the food is functioning as a **lure**. Your intention is to entice the dog to follow the movement of the **lure** and move into a particular position, thereby learning or practicing a task or exercise. When the dog gets to the desired position, the treat can then function as a **working reward**, therefore a **lure** should be soft and small. Examples of using food as a **lure** include things like using a small treat to guide your dog to line up, moving a treat above your dog's head as you ask for a sit, or moving it down and away from the dog as you ask for a down. The use of a **lure** is sometimes called *shaping* or *prompting*. As these terms imply, the handler is doing as much (or perhaps even more) of the work under these circumstances, while the dog is a follower, who is depending upon the movement of the **lure** to lead her.

Visible, Stationary Food as a Marker

Food being used as a **marker** is very similar in nature to the **lure**, in that its purpose is to point out a correct position or location to the dog. But unlike the moving **lure** (the use of which puts the dog in the role of follower) the stationary **marker** functions more as a fixed target. Here, it is up to the dog to take the initiative. It is the dog's responsibility to zero in on and bring her attention to the correct location that is indicated by the static **marker**. The **marker** is often used at the vicinity of the handler's left hip when the dog is in heel position. In Attention Training (Chapter 6), we call this "the *watch* spot." Another common location for the **marker** is at the handler's belt buckle, showing the dog where to aim for a straight front as she approaches her handler during the recall exercise. And as with the **lure**, the motionless **marker** can become a **working reward** once the dog is in the correct position—so, once again, the treat should be of the soft and tiny variety.

As you can see, food can be used in several ways during training. The distinction between these various uses will most likely be quite obvious to your dog. So, to maximize the clarity of your communication, it's important that you keep these differences in mind when you're using food in your training.

Back-Stepping

Always back-step if your dog appears confused or unsure of himself. To "back-step" means going back to something easier or less demanding; such as an easier version or level of a particular exercise. Or, it can mean repeating the same exercise, but changing the mode of training; going back to the helping mode of Wow if your dog struggles with an exercise in Confidence Building, or aborting a test and repeating the exercise in Confidence Building. It is important to be ready to back-step if you *even suspect* that your dog has forgotten something or doesn't understand what you are asking. Never push forward in training an exercise if the dog isn't getting it. Remember, just because your dog knew the exercise yesterday doesn't necessarily mean he isn't confused by it today. We all forget things sometimes! So, if there is any struggle at all, stop immediately and back-step to the previous level or mode where you know you will have a success. This will build confidence for both you and your dog, and strengthen your foundation of mutual trust and respect.

Correction or Back-Step?

Remember, the hard and fast rule is *only give a correction when you're sure that your dog understands what's being asked of him.* If you find yourself wondering whether to give a correction or to back-step, that means you're wondering whether or not he understood you, which means you're not sure. So the choice is clear: back-step!

Using a Training Log

On the right is an example of a page from a training log that records the exercises and the mode (Wow, Confidence, Testing) for each exercise or work segment in a day's training session. Of course, your log could be more or less detailed than the one we show here and could also include things like names of other people and dogs present, number of repetitions, length of session, type of distractions, and so on. Keeping a written record helps keep you honest with yourself about your training progress—and of course it's nice to be able to refer back to it weekly or monthly. At the end of the week or the month, you can use your records to create a pie chart for all of your training over that period of time or you could even break it down further with a separate pie chart for each exercise. You'll be able to see your actual percentages of Wow, Confidence Building, and Testing with pie charts created in this way, using the data right from your own training logs.

Date: afternoon - May 10th - cloudy, cool day

Where: field behind parking lot at the supermarket

Warm Up: ball play and attention

Exercises: recall (W) - smooth
repeat recall (C) - great result!!
work on quick drops (W)

Heeling Segments: forward, about, halt (W)
f, a, h (C) corrected - go back to Wow
f, a, h (W) good!

Stays: (C) - broke sit (lots of distractions)
back-step to onlead-stays (W) (good!)
repeated off-lead stays in different area (W)

Go-Outs & Jumps: (W) note: great sits!

Fun Time: more ball playing - cookies!!

Chapter 5
The Two-Part Foundation

Part I
The Progressive Class

The Foundation of the Bridge

Like all bridges, our bridge must be built on a solid foundation. The two-part foundation for this bridge is a combination of community and partnership. In this chapter and the next, we examine the components of this foundation.

The *community* of people and dogs participating in an obedience class supports learning by example, which for many, is the most efficient way to learn. For years, Marsha has been teaching her students in "Progressive Classes." The Progressive Class is a group of handlers and dogs of differing levels of skill and understanding; such mixed groups benefit both humans and dogs by allowing them to watch and learn from one another. Here, in Chapter 5, we explore the Progressive Class.

The second part of our foundation is the *partnership* between person and dog. A true partnership between each handler and her dog must exist for the high level of communication that is necessary in successful training. In Chapter 6, we offer a method for enriching that partnership using a ten stage program of Attention Training.

Community—The Progressive Class

A community is a group of individuals who share some common characteristics. We often think of a community as a group whose members are helpful to one another in terms of their shared experience and by setting examples of behavior.

Community is a frequently overlooked factor in the learning process. Of course, it is possible to learn a new task by instruction only, but just think about how much faster you learn when you are surrounded by others who are demonstrating how to perform the task.

The world of business recognizes the value of specialized groups that have been termed "Communities of Practice." Do an internet search on this term and you'll find that there are thousands of websites devoted to these purpose-specific communities. A community of practice is more than a group of people who simply share an interest, it is a group of individuals who are engaged in learning together, in sharing knowledge with each other, and in gaining insight and skills from one another. In creating communities of practice, businesses are responding to the findings of social scientists who report that learning is a fundamentally social act. It has been shown that participating in such a community enhances learning and is a great support in realizing one's full potential.

Like humans, dogs are social animals and they too can learn by watching others in their community. The *Progressive Class* is a community that is tailored to the training experience. Instead of segregating students and their dogs by skill level (such as beginner, intermediate, and advanced), a Progressive Class includes a wide range of students. In a Progressive Class, both dogs and humans learn almost as much from one another as they do from the instructor. In a Progressive Class, we come to understand and acknowledge the fact that we all are communicating with each other all the time. The dogs are learning from watching other dogs in the class as well as from being attentive to their human partners' commands and requests. Likewise, the humans are learning from watching their classmates working with their dogs, as well as from listening to the instructor. As handlers, we can learn a lot from simply watching other handlers work with their dogs. We deepen our understanding of dogs in general by watching dogs of differing temperaments, learning styles, and skill levels. The community of a Progressive Class can be enormously helpful to all the participants, canine and human.

Please keep in mind that we do not claim to be presenting scientifically evaluated evidence of social learning in dogs or of the effectiveness of any particular style or method of teaching or training. In the paragraphs that follow, as with many of the ideas in this book, we are simply describing some of the actual experiences that we have had over the years. In presenting the concept of what Marsha has dubbed the *Progressive Class*, it is merely our intention to share with you a description of a learning environment that we have found to be extremely useful in helping dogs and their handlers learn and develop their skills in the obedience exercises.

We realize that the concept of the Progressive Class may be new to many people, and it may be difficult to find a program to attend. If you are an instructor, we strongly suggest that you try this type of class, or if you are a student at a dog school, perhaps you could suggest forming such a class to your instructor. If you are not able to find or create a program, you can accomplish nearly the same thing by organizing a practice group with several friends and their dogs. And if you share your life with more than one dog, they can benefit from watching one another if you work them together.

How Does It Work?

You might be wondering, "How can the instructor plan the lesson if there are beginners as well as intermediate and advanced students all in the same class?" That is a great question! And the answer is surprisingly simple when you realize that each exercise can be performed *at any level*, so the day's lesson focuses on the exercise itself, not on the level of the exercise.

For example, although the dumbbell is not used in the obedience trial ring until the Open level, when taught very gently, even a young puppy can happily learn to hold a dumbbell, and then a bit later, to carry it for a few steps (actual retrieving can come much, much later). It is easy to imagine how watching an Open dog eagerly practicing her retrieve can be immensely helpful for the young puppy who is just learning to hold his dumbbell. The puppy will learn that the dumbbell is an object that older dogs associate with their work and, in listening to the older dog and her handler working together, he will even begin to become familiar with the vocabulary that goes with the dumbbell exercises. Similarly, one could conceive of an instance where an experienced handler (whose own dog is having a minor problem with an advanced exercise) could suddenly gain an important insight into the basics of the exercise while helping or watching another person teach their puppy at a lower level of the same exercise. So you can see that in a Progressive Class, each participant can have a training experience tailored to his or her own level while at the same time benefiting from watching others perform the same or similar exercise at other levels of performance and understanding.

In fact, at times the dogs will even willingly help one another learn. All dogs are social creatures, and dogs who are accustomed to the Progressive Class atmosphere often will pay close attention as their classmates are working.

Watching one another can help the dogs in various ways. One can easily envision the less experienced dogs learning new behaviors from watching more experienced dogs perform. But we also see dogs in this environment who clearly seem to enjoy helping other dogs and demonstrating their skills for their less experienced classmates. In one amazing episode during one of Marsha's classes, we watched as an OTCH dog uttered a grumbling sound as his classmate, a less experienced dog, reached for the wrong scent article during a training exercise. Upon hearing this canine version of a mild scolding, the dog performing the exercise dropped the incorrect article and picked up and retrieved the correct one, to the obvious satisfaction of the dog who had helped him! Of course, that's an extraordinary example, normally the communication between the dogs is a bit more subtle than that, but for those of us who participate regularly in Progressive Classes, the beneficial interaction between the dogs is crystal clear.

The value of social interaction to the learning process is shown in the effectiveness of the training system known as model/rival, or M/R, technique which has become known to many people through comparative psychologist Irene Pepperberg's remarkable work with African Grey parrots. Dr. Pepperberg has used this training technique while studying the cognitive abilities of parrots. She describes the M/R technique, initially designed by German ethologist Dietmar Todt, in which one human "trains" another human in the presence of the parrot. The "trainer" gives reinforcement to the "trainee" using praise and a reward related to the task for a correct performance and a verbal scolding for a wrong response as well as a request to repeat the task. As the parrot watches this interaction, he becomes curious and motivated to learn the task, and indeed learns from watching. Dr. Pepperberg has pointed out that there are many advantages to this type of social stimulation when compared with the more standard training through operant conditioning where the student is simply given positive or negative feedback for his or her performance with no real social interaction.

In any group class, dogs can watch others working. But in the Progressive Class, where dogs of varying skill levels are present, they can observe other dogs who are already quite proficient at exercises to which they themselves are just being introduced. In this way, the Progressive Class creates a learning environment that is similar to the model/rival technique. Having a model performance, during which they observe the enthusiasm of the other dogs and their handlers, and hear the praise and see the rewards that the other dogs are receiving, helps them become excited and motivated to learn the exercise as well.

A good example of using another dog's performance as a model to help create the motivation to learn a new exercise was Shalini's experience while teaching her young Scottish Deerhound to retrieve the dumbbell. She tells the story:

> *My youngest dog, Peri, was eight or nine months old and already understood how to hold and carry the dumbbell. She could skillfully reach for the dumbbell, pick it up and carry it. However, she was having a difficult time moving on to the next level, learning to go after a thrown dumbbell and retrieve it. I asked our two older dogs, a Great Dane and an older Deerhound, both reliable retrievers, to help me teach Peri the exercise. I closed Peri into her x-pen in the living room at home, and as I did so, I explained to her that I wanted her to watch closely as the other dogs practiced their retrieves. I tossed the dumbbell from across the room so that each time it landed near the edge of the x-pen. The two older dogs took turns performing a few retrieves each. Both dogs were enthusiastic about helping, and both performed their retrieves perfectly—sometimes even stopping briefly after picking up the dumbbell to glance at their captive audience, our student Peri in the x-pen, as if to say, "Are you watching this?" After a few retrieves each, it was Peri's turn. I set her up for the exercise in the same place I had with the others, tossed the dumbbell toward the empty x-pen, and sent her to "Get it!" She retrieved like a pro for the first time ever and was obviously very happy with herself. She went on to become a reliable retriever. In fact, she loves carrying and bringing me things so much, that for several years now, it has been her job every evening to collect all the dogs' dinner bowls and bring them to the area where we prepare their meals. She also can be counted on to willingly collect a dropped glove or a toy left in the yard when asked.*

This story, in addition to being a good example of using a model performance to teach and create motivation, shows another aspect of social interaction. Shalini *asked* the older dogs to help show the younger dog how to retrieve. It was pretty obvious that they understood what they were being asked to do and wanted to help. We know plenty of advanced dogs who obviously love their roles as "demo dog" during class or practice, showing off for their less advanced peers. When we become aware of how much interaction can occur between dogs, we may realize that we have been overlooking this important aspect of training. It helps to *ask* our dogs to help their friends and classmates. When you ask an older or more experienced dog to help teach a dog who is learning something new, you are showing your confidence and admiration of her knowledge and abilities. Many dogs who have spent a lot of time learning and perfecting obedience exercises thrive on such compliments.

Preparation for Participating in a Progressive Class

While the Progressive Class allows dogs of all levels of experience to work together, it is important for new dogs to have the basic social skills necessary to participate without causing any disruption to the class. They must also be comfortable walking on lead and be able to sit quietly while others are working. Puppy kindergarten can be a great way for some dogs to gain this basic socialization, but it is not absolutely necessary, especially if the dog is already friendly with other dogs and the dog's handler is experienced with obedience in general and the Progressive Class format in particular. And, of course, the humans taking part in these classes must have a basic understanding of the three modes of training, *Wow*, *Confidence*, and *Testing*. But as long as people who are joining a Progressive Class for the first time are given some introductory instruction on the theory of the three modes of training and the language that's used, handlers of all experience levels (beginner, intermediate, advanced) can easily work together in the same class—and learn from one another while they're at it.

The instructor comes to the Progressive Class with a lesson plan which usually includes a certain set of obedience exercises, with a block of time dedicated to each exercise. Sometimes a class can also include a fun game, such as agility style jumps. And because the class is made up of students who understand the bridging method using the *Wow*, *Confidence*, and *Testing* modes of training, the instructor will often ask for input from the students regarding which mode they would like to be working in during each exercise, and will help the student decide if there is a question about which mode might be best for a dog on that particular day for that particular exercise.

An Example of the Progressive Class

Let's take a look at a typical Progressive Class with a range of skill levels… first let's meet our fictional dogs and their human partners: Trey and Hannah, Rhonda and Andrea, Igor and Madeline, and Salsa and James. They are in the Tuesday evening class. This is a great group, they all enjoy one anothers' company even though they are all so different—or perhaps even because of it! They have become great friends—they bring cake to class when it is someone's birthday, and occasionally will take the dogs out for ice cream after class on a summer evening.

Igor is a six-year-old Miniature Schnauzer who recently earned his OTCH (Obedience Trial Champion). Earning this title requires intelligence, talent, and skill, as well as a lot of hard work and dedication from both partners. The title of OTCH is sort of the canine equivalent of a PhD and not many dogs achieve this distinction. Igor's partner is **Madeline**—she is a corporate executive, she is single, and she thrives on attention to detail, both at work and in her training time with her dogs. When her older dog was nearly ready to retire from competition, Madeline decided to search for a puppy who was bred to be a top performance dog. Her puppy search led her to Igor. Igor's mother is also an OTCH and his sire is an Agility champion. Igor loves to work and shares Madeline's attention to detail. They continue entering obedience trials, always with the goal of achieving HIT (High In Trial), but no matter what the result of a day's competition, they always have a great time. It's easy to see that they love working together. Igor and Madeline are a perfect pair—as well as the role models for the rest of the class!

Rhonda is a three-and-a-half-year-old Afghan Hound and **Andrea** is a graduate student working toward her doctorate in anthropology. Rhonda and Andrea started as beginners in our Progressive Class a couple of years ago. Andrea has a very busy schedule, but she makes it a priority to get to dog school. She says that working with Rhonda helps her relax and keep her sanity. Because Andrea sets such high standards for herself in everything she does, she explored the various dog training programs in the area and decided that our Progressive Classes sounded like a great idea. Rhonda and Andrea earned their CD a few months ago and although they are now focused on their Open-level exercises, Andrea has decided that she doesn't have time to think about competition for a while. So for now, she's just enjoying her training time with Rhonda as well as the enhanced communication in their daily lives that results from their work together.

Trey is a four-year-old Golden Retriever who has earned two legs toward his CDX and his partner, **Hannah,** a recently retired school teacher, is married with two grown children. Hannah and her family are committed "animal people" and have lived with many beloved dogs and cats and other furry creatures over the years. Now Hannah finally has the time in her life to do something just for the fun of it and she is really enjoying her work with Trey, her first obedience dog. Trey and Hannah went through Novice A while taking classes with another school and then, after earning a CD, they joined our Progressive Class. They both caught on quickly to the three modes of training and are developing an even greater partnership. Trey loves the work and has really been coming into his own lately, developing confidence in his understanding of the exercises. The whole family sometimes comes to watch the training classes and are always there to cheer for Trey and Hannah at the trials. Their first two qualifying scores in Open were cause for big family celebrations! Hannah recently entered Trey in an upcoming obedience trial. They are preparing for the competition with the goal of a qualifying score which would give them their third leg, finishing the CDX.

Salsa is a friendly, intelligent, nine-month-old, Shih Tzu-Cocker Spaniel cross and her companion, **James,** is an older gentleman whose wife passed away last year. Salsa has brought a new dimension to James' life. The pair recently graduated from a beginner class at a pet supply chain store. Although James had not planned to continue training after that, his veterinarian saw the potential for these two and suggested that James check out our Progressive Classes. A visit to the class convinced James to sign up and now he and Salsa are the newest members. Because Salsa is a mixed breed dog, they will compete in UKC trials. But titles are not all that important to James. For James and Salsa, the biggest reward of training is the growing understanding of each other that helps make their everyday life together happier and more fun. They both enjoy coming to class and spending time with all their friends at dog school. Salsa is such a cute and happy little dog, and wherever they go—whether it's at shows, practicing at the park, or just going for a walk—she helps James meet people and make new friends.

The Class

The Heeling Exercises

In the Progressive Class format, the heeling exercise can easily be adapted for each pair of participants. Heeling can be broken into segments or done in a whole pattern, and can be practiced on- or off-lead. This can also be a good time to review a few stages of Attention Training (Chapter 6), and, as with all the exercises, heeling can be done in Wow, Confidence, or Testing depending on where individuals are in their own work.

Rhonda's Heeling

Rhonda and Andrea take their turn first. Because their heeling in Wow mode has been free from struggle in their recent training sessions, Andrea decides to try some on-lead heeling segments in Confidence. The instructor suggests a segment of forward/halt/forward/release, and another segment of forward/slow/normal/about-turn/release, both in Confidence.

Andrea does a quick stationary attention exercise to warm-up, with a release to the treat she's holding in the "watch spot" at her left hip. After the warm up, Andrea puts the treats away and then lines up with Rhonda—ready to begin in Confidence mode.

Rhonda notices right away that there is now no visible food in Andrea's hand, and that Andrea is being rather quiet. Rhonda stays close, in good heel position, as they step forward at a normal pace.

Andrea whispers, *"Good!"* quietly complimenting Rhonda on her performance.

Rhonda is performing very nicely until they get to the transition to a slower pace. As Andrea slows down, Rhonda forges ahead just a bit.

Andrea gives her a quiet verbal correction, *"Watch!"* while simultaneously touching her lightly with her hand.

Rhonda is a sensitive, willing worker and adjusts her pace immediately, and Andrea completes the correction by complimenting her, *"That's a better watch—good girl!"* Rhonda performs beautifully for the rest of the segment!

Igor's Heeling

Igor and Madeline go next. After a weekend show that was a bit stressful for both of them, Madeline plans to start with a couple of on-lead heeling segments in Wow, and if that goes well, then bridge the heeling exercise in Confidence. Their on-lead heeling in Wow is fun, Igor performs very nicely, and Madeline releases him with a happy cheer and some cookies!

Since there was no struggle in Wow, they do another on-lead heeling segment, but this time in Confidence Building mode. Madeline hides the treats and they line up to begin.

Igor does very well, watching his partner and trotting along in perfect heel position while Madeline compliments him quietly a couple of times during the exercise. Again, Igor performs the exercise precisely and eagerly—clearly there was no struggle to get it right. So Madeline decides to bridge the heeling exercise a little further—this time off-lead in Confidence—another success!

Going from *on-lead Wow* to *on-lead Confidence* and then to *off-lead Confidence* helps Igor sharpen his performance of the heeling exercise.

Salsa's Heeling

Next, it is Salsa's turn. Salsa and James start by reviewing some exercises from the stages of Attention Training. They take a few minutes to practice the *stationary* attention exercises from stages two and three, as well as the *moving* attention exercises of stages seven and eight (see Chapter 6 - Attention Training). Then they do some heeling segments in Wow with the help of Trey and Hannah who are acting as role models. Salsa and James line up along side Trey and Hannah and together they perform a short heeling sequence of forward, halt, forward, right turn, fast, normal, release. Salsa gets a sense of confidence from Trey's heeling nearby, and she does a great job. What a cutie she is!

Trey's Heeling

And finally, after helping Salsa with her heeling, it is Trey and Hannah's turn to do their own heeling exercise. Hannah chooses to do a few heeling segments in Wow. Trey has been doing very well with his heeling lately and so Hannah's main goal here in choosing Wow is to strengthen their bond rather than to school Trey's performance on this exercise. Hannah holds a visible motivator at the watch spot by her left hip; she's using little tiny pieces of chicken—Trey's favorite treat! Trey is super attentive today throughout his heeling exercises, bouncing and smiling the whole time!

The Retrieve Exercises

There are several obedience exercises that require the retrieval of an object. Dumbbells, gloves, and scent articles (both leather and metal) are all official retrieval objects. Before learning the different retrieval exercises, a dog must learn to "take it" and "hold it" and also to carry the object, so that he learns all about each type of retrieval object. Even young and beginner-level dogs can take part when it's time to practice retrieval exercises in a Progressive Class. When you consider all the different objects, the different modes of training, and the continuum of skill level—from learning about holding a dumbbell to practicing scent discrimination—you can see that there are endless variations on the theme for participants to tailor the retrieval practice to their own needs during the class.

Igor's Retrieve

Igor and Madeline playfully practice scent articles in Wow. She sends him out for the metal article, he goes out, works the pile, sniffing each one, finds the correct one, grabs it and brings it back. Igor sits in front of Madeline and, when asked, gently gives her the article. Then, instead of requiring Igor to do a formal finish, she kneels down and encourages him to jump up and get some big hugs and praises!

Next, as Igor is bringing back the correct leather article, Madeline takes a few gliding steps backward, urging him to come in even faster. He races to her and slides into his front position, Madeline takes a breath, relaxing for just a moment, then takes the article and lets him jump up for a treat. Then she sets him back up in the front position and signals Igor to finish, he does a perfect finish—and they celebrate with cookies! Breaking up the front and finish like this helps keep Igor less likely to anticipate the finish.

Salsa's Retrieve

Salsa is learning to take her dumbbell from James' hand and hold it. Today she held the dumbbell securely and learned how to continue holding it while standing up. What a good girl! Of course this early teaching is all in big-time Wow mode, lots and lots of praise and treats at this level—she's still a puppy. When she has mastered holding her dumbbell, she will progress to walking forward while holding it.

It may be a very long time before she is ready to do much more than that, but for now, watching the older dogs retrieving their dumbbells, articles, and gloves and listening to the verbal cues they are getting from their partners are almost as important as taking her own turn holding the dumbbell. When she is ready to move on, the dumbbell will already be part of her world and the retrieval exercises and their vocabulary will be a familiar part of her work.

Trey's Retrieve

Trey and Hannah choose to do the flat retrieve in Confidence as they prepare for the trial that's coming up later this month. Trey has been performing this exercise very well lately and there has not been any hint of a struggle in getting this one just right for quite some time. He really loves his dumbbell and Trey knows he is good at this exercise so practicing it in Confidence mode today will help sharpen his self-assurance.

Rhonda's Retrieve

Rhonda is very comfortable with carrying her dumbbell and knows how to hold it tightly until asked to release it to her handler. Recently, she has been reliably performing beginner-level retrievals—which we define as going after and retrieving a just-thrown, still-moving dumbbell.

Today they will try the next step in learning about the retrieve. This will require Rhonda to watch patiently while the dumbbell lands and stops moving, waiting until she gets her cue from Andrea before running out to get it. Since this is her first time at this level of the exercise, Rhonda and Andrea are working in Wow.

First, in order to have a happy warm-up, they begin with the throw-it-and-get-it version that Rhonda is familiar with and does so well.

After a couple of these warm-up retrievals, Andrea talks to Rhonda about this new way of retrieving, telling Rhonda that, *"This time, I want you to wait just a little bit—I'll throw it, then we'll wait, then, when I tell you, go get it and bring it back!"* As Andrea explains the slightly new version of the exercise, she is visualizing Rhonda performing it. This helps support Rhonda while she's learning something new.

As Andrea throws the dumbbell, she helps her dog by repeating, *"Wait…"* Rhonda waits for her cue like a pro, and then bounds away after the dumbbell when Andrea sends her after it. And while Rhonda is working, Andrea is continuing her verbal encouragement, *"What a good girl!"* This really helps Rhonda and she brings back her dumbbell with obvious pride in her work and in her new understanding.

Andrea and the instructor are so happy with her performance, they decide to stop on a good note for today. They will continue practicing this new skill often, always in Wow, in the coming days.

The Jumping Exercises

It is easy to imagine all the ways in which jumping exercises can be tailored to each pair of students in the Progressive Class. From the very beginner level of a simple recall over a pole lying on the ground between standards to the advanced directed jumping exercise, there are a huge variety of ways in which we can practice jumping exercises. In order to decide the order of go and how to set up the jumps for the day's training, the instructor will ask which students would like to practice Open-level jumps (high jump and broad jump) and who would like to practice Utility-level jumps (high jump and bar jump). In our example class, the two pairs who choose Open-level jumps will go first and then the other two will practice their jumping with the Utility-level jumps.

Trey's Jumps

Trey and Hannah are going to do the Open jumps in Testing mode. Trey has been practicing the jumping portion of Open level lately and he has been performing the exercises with ease and without hesitation or struggle, so Hannah and the instructor agree that it is a good time to test the exercise, especially since they are getting ready for a trial in the near future.

Of course, in planning to work in the Testing mode, Hannah must first decide at what point during the exercise she will do the jackpot. She plans to do the jackpot just as they are lined up and ready for the broad jump exercise.

They begin the test with the retrieve over the high jump. As Trey lines up with Hannah, he happily notices that she is holding the dumbbell; he sure loves to retrieve! He also notices that Hannah is being a little quiet, so it's already clear to him that they're not training in Wow.

She throws the dumbbell and he eagerly waits for her cue. When she gives the command, he bounds over the high jump, grabs the dumbbell, and as he turns back toward the jump, he realizes that Hannah is watching him in silence. He jumps back over, trots in and sits in a perfect front, gently dropping the dumbbell into her hand when asked. Hannah is still completely silent except for the cue to finish.

Now Trey *knows* this is a test and he's beginning to think that a jackpot might be coming at any time. Hannah and Trey head over to the other side of the ring to line up for the broad jump.

After announcing the exercise and asking if they are ready, the instructor says, *"Leave your dog."*

Instead of giving Trey the cue to wait, Hannah shouts, *"Lotsa cookies!!"* and digs deep into her pocket for a few treats. Trey jumps up, bouncing around in front of Hannah, eagerly eating the cookies.

After a minute or two, Hannah asks Trey to line up again in heel position. Hannah takes a deep breath, relaxes, and nods to the instructor that they're ready to resume the test.

The instructor repeats, *"Leave your dog."*

This time, Hannah signals Trey to wait and leaves him to head toward the jump. From her position facing the broad jump, Hannah calls out, *"Over!"* Trey jumps beautifully, lands and turns to his right and does a perfect front and finish. Great test!

Rhonda's Jumps

Rhonda and Andrea go next. Andrea chooses to do Open jumps in Wow. Because Rhonda is just learning to wait for the just-thrown dumbbell to come to rest in the flat retrieve, Andrea sends Rhonda over the high jump to *"Get it!"* while the dumbbell is still moving. This is a good strategy, Rhonda jumps beautifully, over and back, and brings Andrea the dumbbell with pride in a job well done. Andrea also makes the broad jump fun by throwing a toy past the jump as she sends Rhonda over. Rhonda negotiates the broad jump perfectly, pounces on the toy, squeaking it as she brings it back to Andrea—a fun time for them both!

Igor's Jumps

Igor and Madeline are going to do Utility-level directed jumps in Confidence—this is one of Igor's favorite exercises! But first, Madeline does a quick warm-up in Wow. She sends Igor away and when he reaches the far end of the ring and sits facing her, she tells him to wait and walks out to give him a treat. Besides this little bit of food adding to the fun of Wow, waiting like this helps Igor control his anticipation of Madeline's signal to jump as he has a tendency to be a little over-eager during this exercise. Madeline walks back to her place, relaxes for a moment, then gives Igor the signal to jump the bar jump. He's off like a shot and jumps effortlessly, and when he gets to Madeline, because this is just a fun warm-up in Wow, she releases him to another treat, saying, *"Get it!"* instead of having him perform a front and finish.

Now it's time for Confidence—remember that the Confidence mode of training is characterized by compliments and corrections. Madeline sends Igor away, he flashes across the ring and turns around to sit facing her. Madeline softly praises the sit, she takes a breath to relax for just a moment, and then signals for the high jump.

As expected, he jumps beautifully and is so quick that it seems as if he's almost instantly sitting in a perfect front, toe-to-toe with Madeline, waiting for the signal to finish. His finish is a tiny bit crooked, but it was a good try.

Madeline reaches down and gives Igor a soft nudge to correct his position, saying, *"Straight—that's better!"*

Igor is such an experienced performer, he knows that this small correction was the result of his getting a little sloppy with his finish. They repeat the exercise with the bar jump. Madeline is a little more generous with her praise after having to give Igor that little bit of a correction. Igor does a great job, and he makes sure that his finish is perfect and precise this time. Their work in Confidence mode is a big success today!

Salsa's Jumps

Salsa is just beginning to learn about jumps, so James and the instructor keep everything fun and exciting for her, of course, doing everything in Wow. They go last so that she can watch the older dogs have fun jumping before it's her turn to go.

Salsa is going to do a straight recall over both the high jump and the bar jump which are set just a couple of inches off the ground. For now, they are just concentrating on simply getting familiar with jumping over small obstacles of any type.

James stands behind the high jump holding Salsa's favorite fuzzy toy and, when he calls her, Salsa races over the tiny jump heading straight for James and the toy. As she reaches him he encourages her to jump up and grab the toy and they play a gentle game of tuggy for a moment.

James asks her to, *"Give,"* and she happily lets go of the toy. James leads Salsa to the little bar jump and they repeat the whole process. After she's jumped the bar jump, played tug, and let go of the toy once more, James gives her the toy and they head back to their seat, Salsa trotting alongside James, proudly carrying her fuzzy toy.

The Recall Exercises

In addition to performing the recall straight or with a drop, or the more advanced signal exercise, the recall can be further customized by including or excluding the front and/or the finish. When working in Wow, the recall exercise can be made more exciting by doing something unexpected—such as turning and pretending to run away after you call your dog. This creates a fun game of chase and can be a great way to motivate a dog who has a tendency to be slow on the recall. There are many ways to get creative with the recall exercise to keep it fresh and exciting. As the recall exercise is one of the *could-save-the-dog's-life-someday* exercises, we should all continue to find ways to inspire our dogs to understand the importance of the recall and to perform it reliably.

Salsa's Recall

Salsa and James, at their beginner level, go with the regular recall in Wow mode. The instructor reminded James to make an obviously visible marker with a bit of food in his hands, centered on his body at his belt buckle, so that Salsa will understand where to focus as she comes in for her front. She is still very young and although she's been doing well with her recalls in coming quickly and straight to James, lately Salsa has been anticipating James' command and starts out across the room before he calls her. To help her with this today, the instructor stands with Salsa, lightly holding her collar until she is called. James calls her, the instructor lets go and Salsa dashes across the ring and sits in a very nice front, James releases her to the food. Yay Salsa!

Trey's Recall

Trey and Hannah are going to do a drop-on-recall in Confidence. Trey starts out well, he and Hannah are standing across the ring from each other and the instructor gives Hannah the signal to call Trey. She calls him, he begins to cross the ring with his usual bounding stride and drops perfectly when Hannah gives the cue. But suddenly he jumps up in anticipation before Hannah calls him.

Trey realizes that, *"Uh oh!"* he shouldn't be standing up yet, and so he just stands there looking at Hannah, waiting for some direction from her.

Hannah walks over to him, correcting Trey by touching him lightly between the shoulder blades with her finger tips while saying, *"Down."* Trey drops back into position and she responds, *"Good down,"* completing the correction with praise. Hannah backs up into her position and calls him to front, *"Trey come!"* She completes the exercise with a happy release, *"Good boy, that's better!"* Because she had to go in to correct Trey during the drop in Confidence Building, Hannah now back-steps, repeating the exercise, but in the schooling mode of Wow.

This time, Hannah schools Trey during the drop by reminding him, *"Good down, wait..."* and going in with a treat, repeating, *"Good down, wait..."* as she hands him a tiny piece of cheese. She walks back to her position and relaxes for a moment, the instructor gives the cue and Hannah calls Trey to front, *"Trey come!"* and he completes the exercise nicely. The instructor suggests that Hannah and Trey continue schooling this exercise in Wow a few more times this week at home, and then go back to Confidence when it's again going smoothly.

Rhonda's Recall

Rhonda and Andrea work on a drop-on-recall in Wow. Helping our dogs in the Wow mode often includes extra body-language in addition to a lot of verbal cues and praise, especially when they are learning a new exercise. So in today's class, when the instructor gives the signal for the drop, Andrea takes a bold step forward and raises her hand as she gives the command, *"Down!"* giving Rhonda this obvious body language as an aid to stop and drop.

Rhonda drops beautifully, Andrea offers praise, *"Good down, wait."*

They both wait for the next signal from the instructor and then Andrea continues, *"Rhonda, come!"*

Rhonda trots in, doing a nice front and finish, a perfect job! For the next few weeks, they will continue to practice this exercise in Wow, gradually lessening the body language until Rhonda is performing the exercise reliably, still in Wow, with the happy praise throughout the exercise, but without the overt physical cues. At that point they'll be ready to try the exercise in the Confidence Building mode of training.

Igor's Recall

Igor has been having a little trouble with his signal exercise lately, and so Madeline chooses to school the signal exercise in Wow. She has been concerned about Igor's recent lack of attention during this silent exercise and so she begins the exercise slowly, one step at a time. She plans to reward

Igor for every correct response by walking in to give him a treat after each signal. Things begin well, but then as she signals Igor for the down, his attention wanders and he looks away just as she gives the signal.

Madeline walks quickly toward Igor and gives him a physical and verbal correction by gently touching his shoulders in a downward motion with the fingertips of her left hand while repeating the signal to down with her right hand and adding the spoken cue, *"Down!"*

Igor does a perfect down immediately!

Madeline rewards him for the down with praise, *"Good down!"* and with a treat and then, completing the correction by bringing him up emotionally, she releases him with a happy cheer, *"Yay! What a good boy! That was much better! Woo hoo!"*

Igor jumps up, hopping and dancing around up on his hind legs, happily scarfing the cookies Madeline offers him.

After this mini-celebration, they perform the signal exercise once more in Wow. Again, with Madeline walking in to reward Igor after each signal. He is right on target this time, every movement is clean and precise, he responds perfectly to each signal!

The Stay Exercises

The sit stay and the down stay can be tailored to fit the individual dog's training requirements in many ways. A puppy or beginner dog can do a stay on lead. As the dog advances in skill, the handler can stand at longer and longer distances from the dog. When a stay is done in Wow mode, the handler can go back to the dog during the stay to reinforce the stay and motivate the dog with a small food and/or verbal reward. And of course, the more advanced dogs' handlers will go out of sight for the stays.

Igor's Stay

Madeline goes out of sight during each of the stay exercises, but she wants to keep this as stress-free as possible for Igor. So, working in Wow, she comes back into the room and feeds Igor a little treat once during the sit and twice during the down, showering him with praise each time and again at the end of each exercise.

Rhonda's Stay

Since Andrea and Rhonda earned their CD, they have been beginning to work on their out-of-sight stays. Today, Andrea is keeping the stay exercise somewhat relaxed, sort of in between the familiar Novice-level stay and the newer Open-level stay. Rather than either standing directly across the ring facing Rhonda (as in the Novice stay) or going completely out of sight (as in the Open stay), Andrea sits on the bench at the side of the room where Rhonda can see her but not actually make direct eye contact with her. This requires Rhonda to hold her stay on her own in a sense, as her handler is not really "participating" in the exercise as she would in the Novice stay, but at the same time Rhonda has the support of knowing that Andrea is nearby.

Trey's Stay

Hannah has been feeling very good about Trey's stays and, like Madeline, she wants to keep it a happy exercise for her dog today. So she also chooses to do the stay exercise in Wow. She goes out of sight, but returns a couple of times with reinforcement and motivational treats for Trey.

Salsa's Stay

James keeps Salsa on lead for the first few moments, but she is having such a good day that James feels secure enough to drop the lead and back away. Salsa is really starting to understand the stay and today is the first time that she is able to complete the entire one-minute stay with James standing a short distance away. Yay! Good girl!

Chapter 6
The Two-Part Foundation

Part II
Attention Training

Building Partnership Through Attention Training

Now that you are familiar with the first part of our bridge's foundation—the Progressive Class—we'd like you to consider the other important aspect of that foundation: the partnership between person and dog. Although this partnership develops naturally with all of our work together, we specifically recommend the use of Attention Training to help build and strengthen this partnership. Attention Training teaches the dog that the most important thing he or she can do while working with his or her human partner is to pay attention. Focused attention helps strengthen the bond between human and dog and can lead to a deeper level of communication—not only in obedience training, but in all aspects of the relationship.

Most of the readers of this book will already be familiar with Attention Training of one type or another. In recent years, obedience training that puts the focus on teaching the dog to keep his attention on his human partner has replaced the old fashioned compulsory obedience training that many of us remember from years ago. There are several popular variations of Attention Training, the main hallmark of a good program being plenty of happy, attentive dogs.

Why It Works

Attention Training is an efficient and effective training tool because *paying attention* means pretty much the same thing for all exercises. Conversely, putting the emphasis on the *perfection of the exercise* means something different for each exercise. Once your dog knows an exercise, the best thing for him to do is to pay attention to you while he is performing it. If he knows the exercise and is paying attention, chances are that he will do a great job. On the other hand, no matter how well he knows the exercise, if he's not paying attention, his performance will suffer. When you and your dog are well versed in Attention Training, it's easy to head off a potential mistake during an exercise by prompting your dog, *"Watch me!"* reminding your dog to pay attention. This will bring the dog back to what he knows he's supposed to be doing, as opposed to picking on one tiny aspect of the exercise, which could confuse the dog and be detrimental to the performance.

Attention Training is more than just a way for your dog to learn how to pay attention. These exercises are meant to be used throughout your dog's lifetime—as a warm-up at the beginning of a practice session, to help relax your dog in a strange place, or as a way of "joining up" with each other right before entering the ring at an obedience trial. A dog who is well-trained in attention has a sense of responsibility regarding his work and is a joy to work with in just about any situation.

Attention is not a head position!

Attention Training does not require your dog to adopt a particular head position. Some dogs can happily tilt their heads up and watch while they are heeling, but many dogs cannot do this comfortably. Keep in mind that conformation differs quite a bit between breeds and types of dogs, and that even within the same breed, there will be differences between individuals.

Your goal in Attention Training is for your dog to pay attention to you while he is working. Dogs have excellent peripheral vision; a dog can have his attention focused on his handler without staring directly at his handler's face. As you work through the stages of Attention Training, you will find that you can easily recognize when your dog is paying attention and when he is distracted, regardless of his head position.

The "Watch Spot"

Although there's no one perfect head position for the dog, it is important for you and your dog to begin each exercise in perfect "heel position," both of you facing forward, your dog sitting beside you, his right shoulder aligned with your left leg. And for these attention exercises—your head and eyes tilted slightly toward him, and a tiny piece of delicious food visible in your left hand, held at the "watch spot" in the vicinity of your left hip. The food held at the watch spot functions at first as a lure, while your dog is learning to "watch," and later as a marker, once he understands how to "watch." (Recall the discussion of lures and markers in Chapter 4.) The watch spot is a general focus point for your dog's attention. As mentioned in the previous section, not many dogs have the conformation to comfortably look up at the handler's face while working, but if the dog keeps the watch spot in his peripheral vision he'll easily remain aware of his handler and his handler's direction and pace. It's important to note that the watch spot can be anywhere from the handler's upper thigh to her waist, and depends on the size of the dog relative to the handler's height. (See the photos for Stage 1.) To simplify the description of the exercises, we'll just refer to the watch spot as near your left hip, but keep in mind that the actual location of your watch spot will depend upon the relative heights of you and your dog.

Food is a Lure, then a Marker

In the descriptions that follow, you'll see that each attention exercise begins with handler and dog in heel position, with the dog sitting and the handler holding the food in the watch spot. During the teaching phase of Attention Training—while the dog is still learning the basic concept of turning his attention to the watch spot—the handler initiates the "watch" at the start of each exercise by lightly touching the dog's nose with the small piece of food, bringing the food back to the watch spot and saying, "*Watch,*" then praising the watch when the dog looks up at the food. Note that, during this phase, the handler is moving the food and it is therefore functioning as a lure—luring the dog's attention to the watch spot. As the dog comes to understand the attention exercises, he will begin to automatically bring his attention to the watch spot as soon as he and his handler have assumed heel position with the food being held in the watch spot. At this point, the handler can begin the exercise by simply holding the food as a marker, praise her dog for offering the watch and proceed with the rest of the exercise.

Responsibility

The exercises of the first four stages are practiced while standing in place, and then, in the fifth through the tenth stages, handler and dog practice the watch while moving—just a couple of steps at first and eventually while heeling. Ultimately, the dog will come to understand that it is his job to keep his attention on the watch spot while in heel position—whether he and his handler are standing still or moving. The dog's understanding of this concept is a milestone in the relationship, it is no longer the handler as teacher and dog as student in the attention exercises. It is now much more of a partnership in which dog and handler each have their own responsibilities. It is the dog's responsibility to watch his handler at all times when in heel position and, in regard to the attention exercises, it is the handler's responsibility to practice the exercises often and for the life of the dog. It is also the handler's responsibility to praise the watch sweetly and with sincerity each time it is offered.

A final note: when your dog attains the level of understanding described above, the use of food for the attention exercises becomes optional and occasionally you'll simply use your hand placed at the watch spot as a marker.

Stages of Attention

Each of the ten stages in the Attention Training program we present here requires a little more from the dog than the preceding step, both in terms of how much he is being asked to do and in the length of his attention span. What follows is a description of each stage with accompanying photos.

A good guideline for working with these attention exercises is to do each exercise five times in a row, twice a day, spending about a week on each exercise until it can be performed with ease, though this time frame will vary with each individual. When your dog understands the exercise and performs it reliably and without struggle, then move on to the exercise for the next stage.

You can practice these exercises any time and anywhere, but for the first three weeks it's probably best to just do the exercises at home, indoors where there are few distractions. About the only prerequisites for these exercises are that your dog is comfortable on lead and has mastered the sit.

Attention Training

Stage 1 – Watch

With the dog sitting at your left side in heel position and a tiny piece of food in your left hand, lightly touch the food to his nose, bring it back up to a position near your left hip and say, *"Watch."* When he looks up at the food, praise him, *"Good watch!"* Keep the food at your hip, lower your body by bending your knees, and say, *"Get it,"* as you let him jump or reach up and take the food. If your dog looks away, repeat the command to, *"Watch,"* while adding a gentle touch. If your dog is tall, you can easily touch him on the head or shoulder. If your dog is very small, you can use a *very* gentle tug on the lead to accomplish the addition of touch. If you repeat the cue without adding the sense of touch, you'll be *double-commanding*, which will only teach your dog to ignore you, or at least to wait for more than one cue before responding. Do this exercise anywhere in the house where the footing is comfortable and secure for both you and your dog. Repeat the exercise about five times in a row, then later in the day, do another set of five repetitions.

Stage 1 – The "watch spot" can be anywhere from your upper thigh to your waist, depending on the size of your dog relative to your own height.

The release to *"Get it!"* will also be different depending on the dog's size.

Stage 2 – Watch with a pivot

This stage is only slightly different from Stage 1. As before, with your dog sitting in heel position, lightly touch the food to his nose, bring it back up to your left hip and say, *"Watch."* When he looks up at the food, praise him, *"Good watch!"* After praising the watch, instead of simply releasing the dog, pivot to your left, step in front of your dog so that you are facing him toe-to-toe, moving the food to the front as you turn (as shown in the photos at right), and then release to *"Get it!"* As with the previous stage, repeat the exercise about five times in a row, then later in the day, do another set of five repetitions.

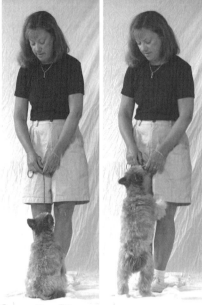

Stage 2 – Pivot to face your dog.   *"Get it!"*

Stage 3 – Add a distraction

It is important to note that by adding a distraction in Stage 3, we are setting the stage for the *possibility* of a correction. Because we do not want to correct a dog who has any confusion about what is being asked of him, you should not move on to this stage until you are as sure as you can be that your dog completely understands your cue to, *"Watch."* So, with that in mind, when your dog is doing reliably well with Stage 2, and you feel sure that he understands the idea of paying attention, it is time to move on to Stage 3. In this stage, we help the dog understand that he is being asked not only to pay attention, but to pay attention no matter what is going on nearby. It is absolutely essential that the handler's intention here is to *help the dog understand* the importance of holding his attention on the watch spot—regardless of distractions. The aim of this exercise (indeed, of all of these exercises) is to support and strengthen his attention, as well as our partnership, so the last thing we would want is for the dog to feel as if he has been tricked into making a mistake or deceived in some way, thereby undermining his trust. This is a good place to remind ourselves of our responsibility to continually cultivate our awareness of the dog's point of view. Keeping his experience of the exercise in mind as you work will help clarify your intention and improve communication.

Attention Training 59

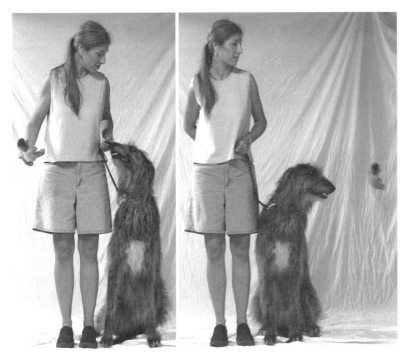

Stage 3 — Toss the object behind you and to your left.

"That's a better watch!"

The exercise for Stage 3 begins in the same way as previous stages... with your dog sitting in heel position, lightly touch the food to his nose, bring it back up to your left hip and say, *"Watch."* When he looks up at the food, praise him, *"Good watch!"* Keep the food at your hip, pivot to your left, step in front of your dog so that you are facing him and then release to, *"Get it!"* However, instead of just doing the usual five repetitions, after you and your dog have completed three successful "watches," the fourth will include a minor distraction. Hold a small object in your right hand (such as a plastic lid, a paper cup, or a small toy), show your dog the food in your left hand, say, *"Watch,"* and as he looks up at you, toss the object behind you and off to your left. (See the series of photos on the previous page.)

It is possible that the first time you try this, the dog will be distracted and look at the object. If he does, give his lead a gentle tug, repeating *"Watch,"* and when he looks back at you, *immediately* give him great praise, a heartfelt, *"Good watch! That's a better watch!"* then step in front toe-to-toe and say, *"Get it."* The importance of using sincere and immediate praise to complete the correction cannot be overstated! Especially in a case such as this where we know there is the possibility of needing to give a correction. So Stage 3 is a chance for your dog to give you proof of his skill in keeping his attention on you, and perhaps, a chance for you to practice the art of the gentle, helpful correction.

Stage 4 – New locations

Continue with the stationary attention exercise as in Stage 3, with the pivot to face your dog and release. The difference in this stage will be to do three of the five repetitions with the distraction as in the previous stage and then two repetitions without the distraction. Once your dog is doing well with the toy-toss type of distraction, it is time to move on to more realistic distractions. If all your work until now has been indoors, you can begin to increase the level of distraction by working outdoors in your own yard—simply being outdoors can be a big distraction at first. As your dog becomes accustomed to working outside, you can increase the distraction again with the presence of other people and dogs, and then finally begin to travel to unfamiliar places— working in a park, a friend's backyard, and other safe locations. Each time you add another level of difficulty to the distractions, be patient and always be ready to back-step a little so that you are not asking too much of your dog. Be aware of your dog's level of sensitivity to new stimuli; what might be a minor distraction for one dog could be quite frightening for another individual. It is important to take your dog's disposition into account in every aspect of training, but this is especially true when adding distractions and working in unfamiliar places. If your dog seems worried about working in a new place, talk to him in a calm, reassuring voice, saying positive things, such as, *"Good boy, this is a safe place. We are safe here."* Try to avoid negatively phrased statements, such as, *"Don't worry,"* or *"No one will hurt you."* The active words in these phrases, "worry" and "hurt," have unpleasant meanings, and we humans tend to subconsciously associate the thoughts and feelings with the words and our dogs can be quite sensitive to this. You can be more supportive of your dog by using positive words which portray the feelings that match your intent.

Stage 4 – Jack keeps his attention on the watch spot while a nearby puppy follows a squeaky toy.

From Stationary Attention to Moving Attention

In Stages 1 through 4, all of our work has been stationary, either in heel position or toe-to-toe. The average length of time for teaching a dog these four stages of stationary attention is about four weeks, although—as with everything we recommend in this book—it depends on how regularly you work with your dog. You'll know your dog is ready to move on to Stage 5 when she easily keeps her attention on you during your work together, even when there are distractions, and when corrections are rarely needed.

Stages 5 and 6 of Attention Training are intermediate steps between the *stationary* attention exercises (Stages 1 – 4) and the *forward-moving* attention exercises (Stages 7 – 10). In Stages 5 and 6, you will be facing your dog while taking a couple of steps backward and the dog will be walking forward and toward you. This makes it easier for the dog to watch while she begins walking and easier for you to keep your dog in full view. These intermediate steps help make a smooth transition from the stationary exercises to the forward movement of heeling exercises.

Stages 7 through 10 are exercises that are performed while you and your dog are moving forward. The value of Attention Training really becomes obvious when we introduce forward motion. When your dog continues to keep her attention on you as you both walk forward, she will naturally be in (or very close to) the correct heel position. And when your dog understands Attention Training, then it is possible to use the simple reminder we discussed earlier,

"*Watch! ...that's a better watch!*" in order to correct for different heeling problems such as lagging behind or forging ahead.

Stage 5 – Backwards heeling

Stage 5 can be described as a "moving watch." It is very similar to Stage 2, in which we pivot to the left and face the dog toe-to-toe and then release to the watch spot or marker—except that, instead of the immediate release, you will take a couple of steps backward, with your dog watching you and moving with you and toward you (see the photo to right). As in all stages of Attention Training, begin with your dog sitting in heel position, lightly touch the food to her nose, bring it back up to your left hip and say, "*Watch.*" When she looks up at the food, keep the food at your hip, praise the watch, "*Good watch!*" then pivot to your left and step in front of your dog so that you are facing her toe-to-toe, move your marker slightly to the front, and take two steps backward with your dog walking toward you, then release her to the marker with a happy, "*Get it!*" As with the previous stages, repeat the exercise about five times in a row, then later in the day, do another set of five repetitions.

Stage 5 – Person moves backward, dog moves forward.

This "backwards heeling" is an ideal way to help the dog keep her attention on the watch spot while introducing the first few steps of movement. The addition of motion can be distracting at first to both the dog and the human. Retaining that focused attention while beginning to move is very important and this arrangement (the dog in front of the handler) allows the dog's attentiveness to be observed much more clearly than when dog is to the side in regular heel position.

Stage 6 – Backwards heeling with turn to forward heel

This exercise is the same as Stage 5, but instead of releasing your dog after taking two or three steps backward, you will pivot back into heel position, with your dog at your left side and continue forward.

Attention Training

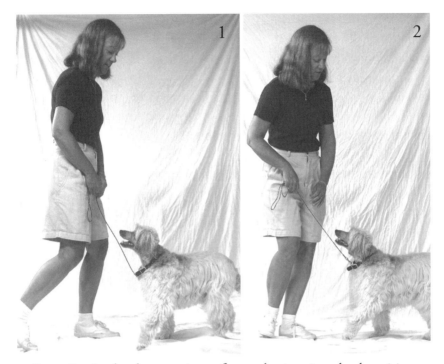

Stage 6 – As the dog continues forward, pivot into heel position. Notice that the marker shifts from the front to the side.

As in all stages of Attention Training, begin with your dog sitting in heel position, lightly bring the food to her nose, raise it back up to your left hip while saying, *"Watch."* When she looks up at the food, keep the food at your hip, praise the watch, *"Good watch!"* Then as in the previous exercise, pivot to your left and step in front of your dog so that you are facing her toe-to-toe, move your marker slightly to the front, and take a couple of steps backward with your dog walking toward you. While continuing to move in the same direction, pivot back to your right, into proper heel position, making sure that your marker moves back to the side and into your regular watch spot, now walking forward with your dog at your side in heel position. Take a few more steps in normal heel position, then release her to the marker, *"Get it!"* Repeat the exercise about five times, then later, do another five repetitions.

Cues

Before we proceed with the description of Stage 7, the first of the forward-moving heel exercises, we'd like to briefly explain two of the cues that we'll be adding from now on, as we describe the moving stages of Attention Training.

"Ready!"

At the start of each exercise in actual competition, the obedience trial judge will name the exercise and ask you if you are ready. For example, the judge will say, *"This is the recall exercise. Are you ready?"* If you are indeed ready, then you will answer, *"Ready!"* and perform the exercise with your dog. We bring this exchange into our training by adding the cue, *"Ready!"* right before the command for any of the forward moving exercises. (By the way, quite often people who are new to obedience competition may not be aware that if they are not actually ready when the judge asks—for example, if your dog's attention is elsewhere—it is acceptable to answer, *"No,"* and quickly finish preparing for the exercise, always being very aware not to unduly hold up the schedule.)

"Heel!"

At the start of any heeling exercise, the handler gives a verbal command to the dog, with or without the addition of the dog's name, as in: *"Jack, heel!"* You can choose to use the word *heel* or any other word that feels comfortable to you and won't be confused with any of the other commands that you use for other exercises. Some words that people use for this cue besides the word *heel* are: *hup, strut, trot, skoh* (a contraction of *let's go*). Limiting it to one syllable is probably best, but the most important thing is to find a word that you are happy with and stick with it so that your meaning is always clear to your dog.

Stage 7 – Two steps, *"Get it!"*

In this stage, begin as always with the dog sitting beside you in heel position, holding the food lightly in the fingers of your left hand, the lead in both hands. Gently touch the food in your left hand to her nose, bring it back up to your left hip and say, *"Watch."* When she looks up at the food, praise her, *"Good watch!"* Keep the food at your hip, but instead of the pivot we add the verbal cue, *"Ready!"* followed by the dog's name and the command to *"Heel!"* Step out with your left foot, take **two steps forward** and then release to the food marker, *"Get it!"* Avoid leaning or moving toward your dog during the release, as this body language will send a negative message to your dog and may cause her to back away. Letting the dog reach to the marker for the food reinforces the watch spot, and has a motivating influence on her.

If your dog's attention begins to wander, gently move the lead upward with your right hand, with the verbal correction and immediate praise, *"Watch! That's a better watch!"*

As always, if you find yourself correcting often, this means that you are moving ahead too quickly. Remember to keep your dog's natural learning pace in mind.

Practice five repetitions of this exercise, once or twice each day until the performance is flowing nicely, your dog is consistently watching you throughout the exercise, and there is no hint of struggle.

"Good watch!"

"Jack, hup!"

"Get it!"

Stage 8 – Heeling in a straight line

The exercise for this stage is the same as Stage 7, but instead of taking only two steps forward, gradually extend the forward motion from the two steps to a long, straight line. Start with three steps, then four steps, and so on, with your dog's attention on you and the watch spot as you both walk forward. End each exercise with the forward-moving release to the food, in order to keep the dog motivated. When practicing heeling exercises, your pace should be brisk and purposeful, walking forward at a speed that is comfortable for both you and your dog. Keep in mind the length of your stride as well as your speed; heeling should have a sense of ease for both partners.

Increase the length of the straight line only when your dog is consistently watching you throughout the exercise, and the performance has a smooth, easy feeling to it. Take notes or otherwise keep track of your progress as you increase the distance of your straight line heeling so that you always know the distance that represents your dog's personal best. This will come in handy in the next stage.

Stage 9 – Add a break-out

In this stage, we introduce the "break-out," a very brief break in the action with an immediate return to heeling and attention. The break-out in Attention Training is not like the jackpot in Testing, nor is it a reward and there is no food or treat given. The purpose of this momentary pause in forward movement is to give the handler the opportunity to practice quickly bringing the dog's attention back to the heeling exercise after a brief interruption.

Stage 9 – Break-out... and return to heeling.

Begin just as in the previous stage, with your dog sitting in heel position, lightly touch the food in your left hand to her nose, bring it back up to your left hip and say, *"Watch."* When she looks up at the food, praise her, *"Good watch!"* Keep the food at the watch spot, and say, *"Ready!"* followed by her name and, *"Heel!"* Step out with your left foot and proceed in a straight line. At some point, *before* you reach your longest distance (from Stage 8), stop and momentarily release your dog with a happy, *"Yay!"* (See photos on the opposite page.) Immediately bring the food back into the watch spot and give her the cue to, *"Watch!"* as you resume moving forward together in heel position—and proceed with the rest of your heeling exercise.

Include the break-out once every two or three times you do the straight line heeling attention exercise. Practicing this interruption will help you and your dog learn to quickly recover your attention in the event of a momentary distraction at any time during your work together.

Stage 10 – Add a circle to the left

In the final stage of our Attention Training, we add circles to the left. Begin just as in the previous stage, with your dog sitting in heel position, lightly touch the food in your left hand to her nose, bring it back up to your left hip and say, *"Watch."* When she looks up at the food, praise her, *"Good watch!"* Keep the food at your hip, and say, *"Ready!"* followed by her name and the cue, *"Heel!"* Step out with your left foot and proceed in a smooth arc to your left, perhaps six or seven steps and then, *"Get it!"* As your dog's attention span increases, increase the number of steps until you are making a complete circle with your dog's attention on you the whole time.

Continuing to use the Stages of Attention

After you and your dog have learned and practiced all ten stages of Attention Training, continue to use these exercises as a warm-up for your regular practice sessions, especially when practicing in unfamiliar locations. Just pick one or two of the stationary exercises and a couple of the moving exercises from the whole ten-stage repertoire to review several times each week. The attention exercises are also great to use as a warm-up right before entering the ring at an obedience trial or match.

Chapter 7

Bridging the Gap

Putting It All Into Action

Now that we have all the pieces, let's pull them together and put it all into action. We have the bridge from training to testing that is the result of using the three modes of training—Wow, Confidence Building, and Testing. And we have described the two parts that make up the foundation of the bridge, the human-dog partnership which can be enhanced by *Attention Training* and a community of other people and dogs in the *Progressive Class* or training group, a social environment that promotes learning.

With all of these components of the training program in mind, let's get an overview of where we're headed. Our goal is to have happy, productive training time with our dogs which results in both better communication with one another in our everyday lives and stress-free, personal best performances in the show ring.

When we are working with our dogs, sometimes we will be in the classroom setting with other people and their dogs and at other times we will be working alone with our dogs at home or in a park or other public place. In our work in the classroom or group setting, we have seen that it is immensely helpful, both to us and to our dogs, to participate in a *Progressive Class*, where all can learn by watching other people and dogs training at many different levels. We are also familiar with *Attention Training* as a way to help our dogs learn to join us in a working partnership.

Remember that for *each exercise*, we can choose to work in any one of the three modes—Wow, Confidence Building, or Testing. While you are teaching your dog a new exercise, you will always be working in Wow mode. After you and your dog really know an exercise, then you will begin the Bridging Process, working a percentage of your training time in each of the different training modes. The percentage of time spent in each training mode is determined by several factors, including things like: how comfortable you both are with the exercise, how recently you've been to a trial or match, or even something as subjective as how both partners are feeling that day.

In Chapter 3, we described an easy way of keeping track of this process using a pie chart to illustrate the percentages of training time you are typically spending in each of the three training modes. The ideal percentages for a dog and handler who have perfected all their exercises are: never less than 70% Wow, about 20% Confidence Building, and no more than 10% Testing. Of course, this is not an exact science, and some dogs may need more of the motivation of Wow to stay sharp and ready to perform at their best. So, as we describe the process of bridging the gap between training and testing, keep in mind these percentages. Before we begin our examples, let's have a quick review of the three modes of training.

Wow - The Motivational Mode

The Wow mode of training helps your dog feel happy and makes his training fun and exciting. Lots of encouragement, using your voice and other rewards, gives your dog the confirmation that he is getting it right. We all like to know when we're doing a good job, and your dog is no different. In Wow, you are assisting your dog on many levels. You are schooling him and helping him understand what you want by giving obvious visual cues. You are rewarding him with a cheerful voice, and helping him in many ways using aids such as food, voice, visible markers, toys, or by breaking up an exercise into segments. In Wow, your dog never has to wonder how he's doing, you are giving him constant feedback in addition to helping him with his performance with obvious body language and encouragement. In short, your dog feels supported and safe in Wow.

The Confidence Building Mode

The Confidence Building mode is the bridge that gets you and your dog from Wow to the Testing stage of training. When you and your dog have been performing a particular exercise beautifully in Wow and there is no feeling of struggle with it, then it's time to move into the Confidence Building stage for that particular exercise.

Working in this training mode builds your dog's confidence in his own ability to perform the exercise without depending on motivation from you. In Confidence, your feedback about his performance doesn't go away (as in the Testing mode) but it becomes more subtle and the timing shifts from during the task to just after it. What does go away is your obvious help, your cheering him on, your encouragement during the exercise. Instead of constantly schooling or helping your dog get it right, you are letting him perform the exercise on his own and then giving him occasional quiet compliments and gentle corrections based on his own performance. Confidence is a distinct mode of training that your dog learns to recognize—when you bridge an exercise using this mode, you use your voice softly and sparingly, and only after he has performed the task. It is quite obviously different from the happy, excited vocal encouragement your dog receives in Wow. Visible markers are very discreet and treats are hidden. Confidence Building helps your dog gain confidence in his own ability and his own understanding of the exercises so that, eventually, he will no longer be distracted or feel worried during those times when your feedback disappears—like in the show ring. He will know that he can do a great job!

The Testing Mode

In many ways, Testing is just like performing with your dog in the show ring. In the Testing mode, like at a trial, the only time you talk to your dog during an exercise is to give him a command. The motive for practicing in the Testing mode is to familiarize both of you with this unnatural style of performing.

If you do not practice this quiet way of working together in your regular training sessions, your dog is going to immediately begin trying to figure out what is wrong when you step into the show ring and suddenly become silent. Not only are the two of you in an unfamiliar ring with a stranger (the judge) watching your every move, there are crowds of strangers and dogs all around, as well as that feeling of nervousness that hangs in the air at a trial. How can your dog even hope to concentrate on his performance if—in addition to dealing with all of these obvious distractions—he's also trying to guess why you have suddenly turned into a silent zombie? For many dogs, wondering why their partner has gone mute is a much bigger concern than worrying about the show surroundings.

But, for your dog, there is more to the Testing mode of training than just getting used to working without hearing your feedback. First, you can continue to send him support and encouragement by thinking of the helping words you would use in Wow and by picturing him performing at his personal best. This will help both of you remain in a calm and confident frame of mind. Another important factor in Testing is the planned jackpot. By planning a break in the action, where you stop and suddenly become the source of *lotsa cookies*, you go a long way in alleviating any tension he might have been feeling while working in silence.

When Testing is done correctly, it doesn't take long for your dog to recognize the quiet of the Testing mode as a potential source of jackpots. This is a great thing, because instead of lagging and wondering what is wrong, your dog will actually learn to look forward to your occasional silent work, because it now means that, at any time, he knows that you might break for a jackpot, so paying attention could really pay off!

Two Examples of the Three Modes of Training in Action

The Novice Recall Exercise & The Heel Free Exercise

Now that we have reviewed the components of the Bridging Process, we will try to bring them to life by looking at a couple of examples, describing both the recall exercise and the heeling exercise in each of the three modes of training. In our illustration of these exercises, we include descriptions from the dog's point of view. Of course, such points of view are fictional, as are the dogs themselves. But we include these descriptions to make an important point—if we, as handlers, can think about, and even identify with, our dog's experience during the training session, it becomes easier to differentiate the three modes of training from one another. For example, as you work in Wow, keep in mind your dog's experience of receiving constant encouragement and schooling from you. When in Confidence, try to feel his experience of becoming more self-reliant. And when Testing an exercise, really put yourself in his place, seeing from his point of view what it's like to work with a nearly silent partner, but being ready for that all-important jackpot at any moment.

In these examples, we reintroduce two of the fictional dogs from our Progressive Class back in Chapter 5:

 The recall exercise will feature Rhonda, the three-and-a-half-year-old Afghan Hound who recently earned her CD, and her partner Andrea.

 The team of the six-year-old Miniature Schnauzer, Igor (an Obedience Trial Champion) and his handler Madeline will serve as our example for the heeling exercise.

The Novice Recall Exercise

The Novice Recall in Wow Mode

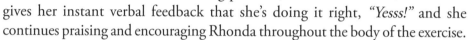

As Andrea begins the recall exercise by calling Rhonda to front, Rhonda knows immediately that they are working in the Wow mode because as she hears the familiar, *"Rhonda, come!"* Andrea's voice is already sending an unmistakable message of happiness and praise! And as Rhonda rises from the sitting position, Andrea gives her instant verbal feedback that she's doing it right, *"Yesss!"* and she continues praising and encouraging Rhonda throughout the body of the exercise.

As she is moving across the ring toward Andrea, Rhonda puts an extra bounce in her stride as she hears her partner exclaim, *"That's right! Yay!"*

Rhonda can easily see the food in Andrea's hand, and she knows that it's a pretty sure bet that she is going to get a food reward for performing this exercise.

As Rhonda approaches ever so slightly off center, Andrea schools her on her front, saying, *"Straight...,"* in a happy, helpful tone of voice, while moving her hands (and food) into the marker position at the midline of her body.

Rhonda adjusts her approach, comes in nicely and sits in a perfect front position.

Andrea drops her hands to her sides, takes a breath, relaxing for a moment. She then brings the tiny piece of food she's holding back up to the marker position and releases Rhonda to the marker with lots of happy praise, *"Great job! Yay!"* (Notice that the food, which first functioned as a *marker*, became a *release reward* at the end of this segment of the exercise.)

After this brief release, Andrea sets her back up in the front position and gives Rhonda the cue to finish. Rhonda walks around Andrea and sits in a nice heel position eagerly looking up at her partner.

Andrea releases her again with lots of praise, *"What a good girl! Yay Rhonda! That was wonderful!!"*

Separating the front from the finish is a good example of schooling in Wow. By emphasizing the separation in this way, Andrea is helping Rhonda understand that the front and the finish are two very different tasks in terms of what is required of the dog. The *front* requires the dog to work somewhat independently—after responding to her handler's command to come, she must find the correct front position on her own and then sit without any further direction from her partner. On the other hand, the *finish* requires the dog to patiently and attentively wait for a cue from her handler before performing the task. It is easy to see why so many dogs make the mistake of going right to finish without waiting for the cue. Between the front the finish, the dog has to make the transition from working independently to waiting for direction. The Wow mode of training is where we can teach an inexperienced dog how to make this transition and also where we can help the more advanced dog keep a sharp distinction between the front and the finish.

The Novice Recall in Confidence Building Mode

Rhonda has been performing the recall in Wow beautifully! This exercise is no longer a struggle for her in Wow and any minor corrections that may be needed are few and far between. Now that Andrea feels sure that her dog understands the exercise, it is time to begin working on the recall exercise in the Confidence Building mode of training.

Instead of the lavish praise and schooling she received from Andrea when working in Wow, here in Confidence Building, Rhonda notices a softer tone of voice than she heard in Wow, but not silence.

Rhonda hears Andrea's praise, spoken in a soft voice—a whispered, *"Yes, that's right."*

As Rhonda approaches Andrea, she hears a quiet helping word, *"Straight,"* but, unlike in Wow, where Andrea's hands move to the middle to provide a marker, Andrea's hands remain at her sides. Rhonda appreciates the subtle assistance of Andrea's quiet voice, it helps her feel confident that she's pleasing Andrea even though she is not receiving constant feedback.

Rhonda sits nicely in front of Andrea, and hears her partner say, *"Good."* Andrea relaxes for a moment and then cues Rhonda to finish. Rhonda goes around Andrea and sits in heel position, again Andrea pauses and then, with the exercise complete, she releases her with treats and praise, *"Good girl! Yay!"*

As the name of this training mode implies, subtle compliments and corrections help build the dog's self-confidence and show her that she can perform the recall exercise on her own. In this mode of training, the handler is helping to build her dog's confidence in her own skill and knowledge of the exercise.

The Novice Recall in Testing Mode

This is, literally, where Andrea and Rhonda will put it all to the test. When Andrea can rely on Rhonda's recall exercise to be nearly flawless in Confidence Building, when they have really mastered every portion of it—including the front and finish—then they are ready to try the recall exercise in Testing mode.

Before beginning a test, Andrea must plan ahead in two ways. First, she must know what her accepted level of performance will be and, second, she must plan the timing of the jackpot. Andrea has limited practice time these days and she's not looking for perfection, so Andrea and the instructor decide that they will only abort the test in the case of a mistake that would result in a non-qualifying score during a competition. And as for the timing of the jackpot, Andrea decides to break-out for the jackpot instead of giving the signal for the finish.

The exercise begins with Rhonda sitting across the ring from Andrea. Andrea has her hands at her sides and she is being very careful not to give any signals to Rhonda that she would not give in the show ring. Just as in a real competition, the command, *"Rhonda, come!"* is the only thing Rhonda hears from her partner.

As Rhonda bounds across the ring, Andrea's hands remain still and hanging at her sides, she will not give Rhonda even a hint of a marker, and there is no food in her hands. One thing Andrea can do in Testing though, is to *think* the helping word or words that she might say when she's working in Wow… she *mentally* says, *"Yesss! That's right!"* as she flashes Rhonda a smile—this gives Rhonda the mental picture of the happy way Andrea helps her in Wow.

Rhonda has come across the ring and is sitting in front of Andrea. Her front is a bit off, just slightly to the left of center, but this is a test and Rhonda put in a good, happy effort, so Andrea overlooks this little mistake.

Rhonda is ready and waiting for Andrea to give her the signal to finish—when suddenly, instead of a cue to finish, Andrea lets out a happy, *"Lotsa cookies!"* and reaches for the hidden treats. She makes it really fun for Rhonda by taking some time out of the exercise at this point so that they can share a fun little break with a nice treat.

Once Rhonda has finished enjoying the cookies, Andrea sets her back up in her front position, she takes a deep breath, lets it out, Rhonda realizes that she's working again, and now she feels fresh and relaxed and happy to get back to her job. Andrea resumes Testing mode and continues with her signal to finish, *just as if there had never been a break in the action.* She is careful to stay in Testing mode until the exercise is complete, with Rhonda sitting at her side in heel position.

The instructor announces, *"Exercise finished!"*

Now she can release Rhonda with a happy voice, it's time to reward her for a job well done—just as she would after a performance in the show ring.

Some Notes on Testing

Aborting the Test—If Rhonda had *not* made a good effort, let's say, for instance, she had skipped the front completely and had just run past Andrea and went right to finish. Then Andrea would have aborted the test, gently corrected Rhonda, praising her for the corrected front and then repeated the exercise in Confidence Building mode, and remained in Confidence, leaving any further Testing of the recall for another training session.

The Jackpot—Always remember that the jackpot is *not* a reward for what your dog has performed so far, it is a break in the action, and it is separate from the performance. By planning the timing of the jackpot ahead of time, you guarantee that separation.

Summary of the Three Modes of Training for the Recall Exercise

Recall	Wow	Confidence	Testing
Intention	To help your dog with constant praise and schooling, to motivate him *while he's working*, to provide continual feedback about his performance.	To allow your dog to work on his own, while occasionally offering quiet compliments and gentle corrections *as he completes a task*.	To be your dog's silent partner while maintaining a strong thought-connection.
Voice	Happy, frequent help, lots of feedback. Instructive, helping words, spoken clearly and encouragingly, such as, *Wait, Straight, Good*.	Infrequent, quiet feedback. Only occasional helping words, spoken very softly, such as, *Straight, Good*.	No voice except command to *Come* and cue to *Finish* and during planned jackpot.
Food	Visible during exercise, use freely as working rewards.	Hidden in hand or pocket. Occasional reward at end of an exercise.	Only in planned jackpot and when finished exercise.
Markers	Hands up to midline of body.	None.	None.

The Heel Free Exercise

The description of the heeling exercise in all three modes of training features Igor and Madeline performing a shortened pattern of:

Forward–Halt–Forward–About Turn–
Slow–Halt–Release.

This simplified pattern, with a forward-to-halt transition, a halt-to-forward transition, a turn, and a pace change, has the elements we'll need to demonstrate the differences between the three modes, while still keeping the descriptions brief and to the point. After reading these examples, you should have no trouble applying the different modes to all the elements in a full heeling pattern.

In our description of the heeling exercise, we include a helper who plays the role of the judge, calling out the pattern for Madeline and Igor. If you are alone with your dog during a practice session in Wow or Confidence Building mode, simply use the command words that you would normally use. But in the Testing mode, it is important to either enlist a friend or family member who can call the pattern for you or else prepare a tape recording to simulate the presence of a judge so that you can reproduce the show ring atmosphere as closely as possible.

A strong foundation in Attention Training (Chapter 6) is always helpful in working with our dogs and this is especially true in the heeling exercises. By starting off with an attention-style warm-up, we can re-establish our working connection with our dogs, or "join up" with them, at the beginning of each heeling sequence. In the following examples of off-lead heeling in each of the three modes of training, our fictional team reinforces their partnership in this way at the start of each exercise.

The Heel Free Exercise in Wow Mode

In the Wow mode of training, lots of schooling and generous encouragement help the dog improve his understanding of the exercises and the constant feedback lets him know exactly how he's doing all the time. Happy praise when he's right, and gentle corrections when he's wrong result in his feeling great about himself and his performance. In this motivational mode, Madeline's constant feedback means that Igor is always sure of how he's doing, there's never any guessing on his part about whether he's pleasing her or not.

Madeline prepares for the heeling exercise with Igor sitting beside her in heel position. Before the helper asks if they are ready to begin, Madeline and Igor join up with a quick attention exercise. With a tiny piece of food curled in the fingers of her left hand, Madeline lightly touches the food to Igor's nose, brings it back up to her left hip and says, *"Watch."*

He knows for sure he's doing well because when he responds to her cue by looking up at her and the food, he hears her happy praise, *"Good watch!"*

The helper announces the heel exercise and asks, *"Are you ready?"*

Igor can see the tiny piece of food in his partner's hand as he eagerly waits for the command.

He hears Madeline answer, *"Ready!"*

The helper calls out, *"Forward."*

Madeline says happily, *"Igor, heel!"*

As the two of them step forward together at a brisk pace, his attention is on her—and her attention is on him.

She encourages the feeling of partnership between the two of them with a cheerful, *"Yesss!"* all the while holding a mental picture of Igor sticking close to her in perfect heel position.

As they approach the location of the first halt, their helper calls the command, *"Halt."*

Madeline helps Igor by making the last step before the halt just a little slower and by bringing her shoulders back a bit as she steps into the standing position of the halt. This accentuated body language gives Igor some extra help with the halt.

As they come to a halt, Igor hears Madeline's happy voice, *"Sit straight, good sit!"*

Their helper continues, *"Forward."*

Madeline looks at Igor with a happy smile and says, *"Igor, heel!"*

And once again, as they move forward together, Igor hears another motivating phrase from his partner, *"Yes, that's right!"* which helps him enjoy their work together as a team and lets him know that he's pleasing her.

They both hear the helper call, *"About turn."*

Madeline steps into the turn while keeping her feet close together, making the turn nice and tight. She helps Igor by giving a focus point with the verbal cue to *"Watch,"* just like in the stationary Attention Training. This focus point reminds him to keep his attention on her, and on the food she's holding at the watch spot, as they make the turn.

"Slow," calls the helper.

Madeline leans back with her shoulders and says, *"S-l-o-w... good slow,"* drawing out her words as she slows her stride.

Igor is confident that he understands the command, *"Slow,"* and matches Madeline's slower steps and her extra cues help him feel even more sure of himself.

As they arrive at the corner of the ring and the end of this shortened heel pattern, the helper calls, *"Halt."*

Madeline takes two slower steps and then stops—Igor stops and sits right beside her.

Their helper announces, *"Exercise finished!"*

Madeline bends her knees a bit, offering Igor the food she's been holding in the watch spot throughout the exercise and releases him to the treat, *"Yaaaay!! Get it! Good boy, Igor!!"* reinforcing his attention by feeding him right from the watch position at her hip.

The Heel Free Exercise in Confidence Building Mode

With Igor sitting beside Madeline in heel position, the two of them are ready to begin the heel exercise. So far, everything is just like in Wow, except Madeline keeps the food hidden and will only use it as an occasional reward for a job well done.

She bends down slightly and says, *"Watch,"* getting Igor's attention.

He looks up and watches her as she resumes her standing position. Igor responds to her cue by looking at her eagerly, his attention on the watch spot. When Madeline smiles back at him but says nothing, he realizes that he will be getting less feedback and will have to take some responsibility for his performance.

The helper announces the heel exercise and asks, *"Are you ready?"*

As he waits for her command, Igor looks up at Madeline's hand in the watch spot. He has long experience in knowing that it is his responsibility to maintain his "watch" on this marker. With his attention on the watch spot, he also notices that there is no food in Madeline's hand, he knows that this is further evidence that they are not working in Wow.

Madeline answers the helper's question, saying, *"Ready!"*

The helper cues them, *"Forward."*

Madeline begins, *"Igor, heel!"*

As the two of them step out together, their attention is on each other.

Madeline quietly supports Igor's self-confidence with a whispered, *"Good!"* and just like in Wow, she continues to hold a mental picture of Igor sticking close to her in that perfect heel position.

Nearing the position of the first halt, Madeline helps Igor just as she does in Wow—by slowing her step just a bit and leaning her shoulders back.

The helper calls out, *"Halt."*

They take two more steps and come to a halt, Igor sitting perfectly at her side. Madeline slips him a tiny treat as a reward for a great halt.

She whispers, *"Good sit!"*

He accepts the tiny reward and quickly swallows the treat, thinking about what will come next.

Igor hears the helper say, *"Forward."*

As expected, Madeline's voice continues, *"Igor, heel!"*

They move forward together and Madeline thinks about what a great job Igor is doing, sending him a warm feeling of appreciation. Igor continues along in heel position confident that he's doing a great job.

Madeline and Igor hear the next call, *"About turn."*

Igor keeps his shoulder right next to Madeline's leg as they both turn to the right and step out briskly together in the new direction.

Their helper calls, *"Slow...."*

Madeline slows down and Igor responds appropriately, slowing his pace to match his partner's stride.

As they are about to arrive at the corner of the ring, the helper says, *"Halt."*

They take a couple of steps and come to a halt.

The helper calls, *"Exercise finished!"*

Madeline releases Igor with a happy, *"Woo hoo!! Yaaay Igor!! Get it!"* offering him a few little pieces of cheese as a release reward at the end of the exercise.

The Heel Free Exercise in Testing Mode

The Testing mode always includes at least one jackpot—a planned breakout to ease the tension of working in silence. So before beginning the heel pattern, Madeline thinks about it and decides to break for the jackpot after the first halt. This means that when the helper calls for them to proceed forward from the halt, Madeline will respond instead by breaking for a jackpot. After Igor enjoys his treat and the break in the action, she will get him back into heel position and they will proceed with the rest of the exercise.

In addition to planning the jackpot, another thing Madeline decides before beginning the test is the level of performance that she expects from Igor today. Madeline and Igor are quite a team; they both enjoy being top performers. Igor recently earned his OTCH and when the two of them participate in a trial, they are there to win. Today they are preparing for a trial and Madeline knows that Igor has really been in top form lately. She decides to go ahead and test for a nearly perfect score. In other words, if Igor were to make more than one or two half-point errors, Madeline would abort the test and make a correction, going back into Confidence Building for the remainder of the exercise.

Madeline and Igor line up in heel position, ready to begin their test of this short heel pattern.

Just like in the other modes, Madeline leans down toward Igor and says, *"Watch,"* and Igor looks up at her attentively.

The helper states that this is the heeling exercise and asks, *"Are you ready?"*

Madeline replies, *"Ready!"*

The helper says, *"Forward."*

And Madeline continues with, *"Igor, heel!"*

As the two of them begin the heel pattern, Igor notices that his partner is being very quiet and that there's no visible food in her hand. It is at this point that many dogs who are not accustomed to the three modes of training start to get nervous about their performance or wonder why their partner is being so unresponsive. Many dogs will start to lag as they have these unsettling thoughts. But Igor has been working in the three modes of training for years and as soon as he notices Madeline's silence and lack of food and feedback, he recognizes the Testing mode, and that means he knows that there is a good chance that there will be a fun break-out (and food!) at any moment.

As they proceed forward, Madeline looks at Igor's nice position beside her and *thinks* to him, *"Good!"* This helps her keep a mental bond with her partner and his performance. And just like in the Wow and Confidence Building modes, she continues to hold a mental picture of Igor keeping very close to her in his usual perfect heel position as they walk briskly forward side by side.

The helper calls out, *"Halt."*

They take two more steps and then come to a halt in silence. Igor sits by Madeline's side and looks up at her.

The helper says, *"Forward."*

But instead of stepping forward, Madeline responds with a cheerful, *"Lotsa cookies!"* and reaches for the hidden food.

Igor leaps up and dances around on his hind feet, eagerly taking several little cookies and joining his partner in the playful break from the exercise.

As soon as he's enjoyed his treat and the break in the action, Madeline calls Igor back to heel position. She takes a deep breath, relaxes her shoulders and nods to the helper that they're ready to resume the exercise. Igor looks up at his partner feeling relaxed and good about his work.

Then, just as if there had never been a break in the exercise, the helper continues, *"Forward."*

Madeline says, *"Igor, heel!"* and silently, they walk forward together.

As they reach the end of the ring, the helper calls, *"About turn."*

They do a nice turn and then hear from the helper, *"Slow...."*

Madeline slows down and Igor slows his pace immediately. Igor looks up at Madeline and they exchange a knowing glance, they're a great team!

As they near the corner of the ring, the helper gives the cue for the final halt.

Madeline and Igor halt in perfect unison.

Their helper says, *"Exercise finished!"*

Madeline releases Igor, *"What a good boy! Let's go get some cookies!"* And just like at a trial, they leave the ring on their way to get some well-deserved treats!

Summary of the Three Modes of Training for the Heeling Exercise

Heeling	Wow	Confidence	Testing
Intention	To help your dog with constant praise and schooling, to motivate him *while he's working*, to provide continual feedback about his performance.	To allow your dog to work on his own, while occasionally offering quiet compliments and gentle corrections *as he completes a task.*	To be your dog's silent partner while maintaining a strong thought-connection.
Voice	Happy, frequent help, lots of feedback. Instructive, helping words, spoken encouragingly, such as, *Slow, Fast, Sit Straight, Good.*	Infrequent, quiet feedback. Only occasional helping words, spoken very softly, such as, *Straight, Good.*	Only for commands and during planned jackpots.
Food	Visible during exercise, use freely as working rewards.	Hidden in hand or pocket. Occasional reward at end of an exercise.	Only in planned jackpots and when finished exercise.
Markers	Left hand in the "watch" spot.	None.	None.

Conclusion

We all love our dogs and want them to be as healthy and as happy as possible, we want to spend our days in their company, and we want them to enjoy our companionship as much as we do theirs. Obedience training can contribute to the fulfillment of these desires by providing an opportunity for exercise and teamwork and for deepening our friendship by spending time together learning and polishing our skills. The Bridging Process in particular is a format that encourages us to not only learn and perform together, but to align our thinking with one another in order to grow in confidence and to really pay attention to one another, creating a framework within which we can fully experience our dogs and they us.

One of the most important things to remember about the Bridging Process is that, while all three modes are equally important in creating a confident and capable dog-and-handler team, the *percentages of training time* spent in each one are far from equal. These percentages change and evolve as we're learning and practicing new exercises. The imaginary wall of training "bricks" that is described in Chapter 2 is an effective way of visualizing how each of the three modes of training influences the results of training. Imagine that the bricks represent the degree of support a dog receives from her handler in each mode of training. Practicing an exercise in Wow adds two training bricks to the wall, working in Confidence Building adds only one brick, while Testing an exercise removes a brick from the wall.

For an experienced team that has polished all of the exercises in their repertoire, a good approximation for the ideal percentages of time spent in each mode is at least 70% in Wow and no more than 20% in Confidence or 10% in Testing. The pie chart illustrates these percentages and makes clear that the majority of our training time is best spent in Wow. It also is a good reminder to continually circle around the chart—from Wow to Confidence Building to Testing and back again.

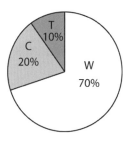

Everything we have discussed can be used toward the goal of achieving a great performance in practice and duplicating it in the obedience trial ring. But there is a more fundamental message underlying it all—throughout this book, from the three modes of training to the progressive class, from the stages of attention to pie charts—that message is one of strengthening the relationships we, as handlers and caregivers, have with our dogs. As noted in the introduction—we believe that it's more rewarding to show to train than to train to show. Whether or not we ever participate in competition, learning and practicing the obedience exercises together helps us create a special bond with our dogs—a bond based on love and friendship, on shared experiences and teamwork, and on all those hours spent working together, thinking together, and communicating with one another.

Marsha sums it all up:

> *I'm writing this while on vacation in the Outer Banks of North Carolina. We're sitting on the sun porch, and my three obedience trained dogs are enjoying the afternoon with me. The majority of this book is about preparing for competition—but the true benefits of the book are illustrated by what I'm doing right now. Due to my hours of training and competing with these guys, we have an unbelievable bond. I'm sure we wouldn't share this amount of understanding, love, and trust had we not been "in the trenches" training and showing together. Not only are my dogs well-behaved, welcome guests in a rental beach house, they are no trouble to care for and truly enhance my vacation, rather than complicate it. Training is:* Dog Talent + Human Talent + Time. *As I sit here with my gang, I know in my heart that it's time well spent.*

Glossary

AKC—the *American Kennel Club* is an organization that maintains a registry of purebred dogs, sponsors dog competitions held by licensed and member clubs, and acknowledges achievement by awarding titles within the various categories of competition. Types of sanctioned AKC competitive events include: Conformation, Obedience, Agility, Rally, Herding, Tracking, Lure Coursing, Hunting Tests, and others.

back-step—to go back to an earlier or easier version of an exercise during a training session, or to repeat the same exercise but change to a more supportive mode of training. Examples: to go from off-lead heeling back to on-lead heeling or to go from Confidence Building back to Wow. A handler should *back-step* whenever the dog makes an error, hesitates, or appears confused.

double-command—the exact repetition of a cue after not getting a response from the dog. A *double-command* is detrimental to one's training progress as it will teach the dog to ignore his handler or to wait for more than one cue before responding. However, by repeating the cue along with the addition of a touch, the handler is emphasizing and clarifying the command, thereby making a correction that helps the dog understand what is required of him. Example: the dog does not respond to the handler's verbal cue, "*Watch.*" If the handler simply repeats the verbal cue, she is giving a *double-command*. If instead she touches the dog's shoulder while repeating the verbal cue (and then praises the dog's response), she is giving a correction.

empathy—identification with or vicarious experience of the feelings, thoughts, or viewpoint of another.

finish—when the dog returns to heel position after coming to the handler, as in the *recall* exercise. The dog may either go clockwise around the handler or pivot to his left and step back into heel position. Both variations are permissible. In competition, the handler may use either a verbal command or a hand signal to cue her dog to finish, but not both.

front—the dog's sitting position, centered in front of and facing the handler after coming to her, as in the *recall* exercise. In a correct *front*, the dog is sitting straight and is close enough for the handler to be able to touch his head without having to stretch or step forward.

Glossary

go-out—in the *directed jumping* exercise (at the *Utility* level), the dog is cued to leave the handler in order to *go out* to the opposite side of the ring where he turns to face the handler, sits, and waits to be directed to jump either the bar jump or the high jump.

heel position—position in which the dog is at the handler's left side, facing the same direction that the handler is facing, with the area from the dog's head to his shoulder aligned with the handler's left hip. The dog can be standing, sitting, lying down, or moving while in *heel position*. The dog should be reasonably close to, but not touching or crowding his handler.

HIT—*High In Trial*, an award given to the highest scoring dog at an obedience trial.

lure—food, toy, or hand which is being moved in a particular direction in order to encourage the dog to focus on and follow the movement of the object, thereby learning or practicing a task or exercise. With the use of a *lure*, it is the handler who has the primary responsibility, as the handler is assuming the role of the leader and the dog is in the role of a follower—compare to *marker*.

marker—food, toy, or hand held in a stationary position as a target on which the dog can focus his attention. A well-trained dog will have learned that it is his responsibility to focus his attention on the location indicated by the static *marker*. Therefore, with the use of a *marker*, it is the dog who has primary responsibility—compare to *lure*.

non-qualifying score—a score earned by a dog in an AKC or UKC obedience competition that falls short of qualifying the dog for a "leg" toward an obedience title. A *non-qualifying score* is a total score of less than 170 points (out of a possible 200 points) in an obedience class or a score in which the dog fails to earn more than 50% of the available points in any one of the required exercises. The judge deducts points for errors or misbehavior by the dog or for the handler's inability to control the dog while in the ring.

Novice—the lowest level of obedience competition in which the dog can earn a title. At the *Novice* level, the dog earns the title Companion Dog (CD) by earning three qualifying scores, or "legs," as certified by three different AKC judges in *Novice* classes at three official AKC obedience trials. With slightly different requirements, the UKC awards the title United Companion Dog (UCD) to a dog earning three qualifying scores, or "legs," as certified by at least two different UKC judges in *Novice* classes at three official UKC obedience trials.

Open—the second level of obedience competition in which the dog can earn a title. At the *Open* level, the dog earns the title Companion Dog Excellent (CDX) by earning three qualifying scores, or "legs," as certified by three different AKC judges in *Open* classes at three official AKC obedience trials. With slightly different requirements, the UKC awards the title United Companion Dog Excellent (UCDX) to a dog earning three qualifying scores, or "legs," as certified by at least two different UKC judges in *Open* classes at three official UKC obedience trials.

Open jumps—the jumps that are required at the *Open* level of obedience competition are: the *high jump*, a wall of solid, white boards held between white standards and the *broad jump*, white boards arranged on the ground to form a flat, ground-level obstacle. The associated exercises are: 1. the *retrieve over high jump*, in which the handler throws a dumbbell over the jump and sends the dog out to retrieve it (the dog jumps the obstacle both on the way out to get the dumbbell and also when returning while carrying the dumbbell), and 2. the *broad jump*, in which the handler leaves the dog sitting in a position facing the obstacle and walks to a position next to the jump, the handler then cues the dog to jump the obstacle and come to *front* and then to *finish*.

operant conditioning—in the fields of Psychology and Behaviorism, the system of changing (*conditioning*) the behavior of a subject (*operant*) by means of positive and negative reinforcements as well as positive and negative punishments. In this context, *positive* simply means that something is added and *negative* means that something is taken away. A *reinforcement* is something that perpetuates a behavior and a *punishment* is something that causes the subject to be less likely to repeat the behavior. Therefore, a positive reinforcement is the addition of desired consequence. A negative reinforcement is the taking away of an undesired consequence. A positive punishment is the addition of an undesired consequence. A negative punishment is the taking away of a desired consequence.

OTCH—the AKC title *Obedience Trial Champion* is awarded to dogs who have first earned the Utility Dog (UD) title and then earned a total of 100 championship points. Championship points are earned by achieving particular placements (first, second, third, or fourth) in a prescribed number of both Open and Utility level classes and are awarded according to a point schedule that takes into account the number of dogs competing in the classes.

qualifying score—the score earned by a dog in an AKC or UKC obedience competition that qualifies the dog for a "leg" toward an obedience title. To earn a *qualifying score*, the dog must have a total score of at least 170 points (out of a possible 200 points) in an obedience class and earn more than 50% of the maximum number of points assigned to each of the required exercises.

telepathy—literally, "*perceiving over a distance.*" The combination of *tele-* which means "*transmission over a distance*" and *-pathy* meaning "*feeling, perceiving.*" *Telepathy* is the direct perception of thoughts, feelings, or images by means other than the usual physical senses, such as hearing, seeing, or touching.

UKC—the *United Kennel Club* is an organization that maintains a registry of purebred and mixed breed dogs, sponsors dog competitions, and acknowledges achievement by awarding titles within the various categories of competition. The UKC places equal emphasis on performance and conformation. Types of sanctioned UKC competitive events include: Conformation, Obedience, Agility, Coonhound Field Trials, Hunting Tests, Weight Pull events, and others.

Utility—the third level of obedience competition in which the dog can earn a title. At the *Utility* level, the dog earns the title Utility Dog (UD) by earning three qualifying scores, or "legs," as certified by three different AKC judges in *Utility* classes at three official AKC obedience trials. With slightly different requirements, the UKC awards the title United Utility Dog (UUD) to a dog earning three qualifying scores, or "legs," as certified by at least two different UKC judges in *Utility* classes at three official UKC obedience trials.

Utility jumps—the jumps that are required at the *Utility* level of obedience competition are: the *high jump*, a wall of solid, white boards held between white standards and the *bar jump*, a single black and white striped bar held between white standards. The associated exercise is the *directed jumping* exercise, in which the handler cues the dog to *go out* to the opposite side of the ring where he turns his attention toward his handler, sits down, and waits to be directed to jump either the bar jump or the high jump. The judge then calls out either, *"Bar,"* or, *"High,"* thereby signalling the handler to cue her dog to jump the appropriate obstacle and return to *front* and then to *finish*. The entire process is then repeated with the dog being directed to jump the other obstacle.

watch spot—the *marker* (food or hand) that is held in a stationary position in the vicinity of the handler's left hip as a target on which the dog can focus his attention while working. Especially as in Attention Training.

About the Authors

Marsha Smith

Marsha was born with a love of dogs and training. As a small child, long before she was able to have dogs of her own, she spent hours "training" her stuffed animals. Today, more than one hundred dogs and their handlers attend Marsha's obedience school, Canine Workout, each week. Noted as one of the "top trainers on the east coast" by a popular obedience magazine, Marsha loves working with dogs and handlers of all levels and delights in each student's personal best performance—whether it earns a qualifying score, a title, a championship, or simply a genuine smile of satisfaction.

Canine Workout, located near Chadds Ford, Pennsylvania, is known for producing successful competitors, with students earning numerous titles each year. Handlers and dogs from Canine Workout accounted for five of the 101 AKC Obedience Trial Championships that were awarded in the United States during the year 2001.

Marsha's life-long interest in teaching led her to major in early childhood education in college which broadened her understanding of teaching methods and education theory. She has been a National Association of Dog Obedience Instructors (NADOI) endorsed instructor since 1987 and has written articles on training for obedience and breed-specific magazines.

Marsha's unique style of training attracts many students who are devoted competitors, as well as those who have no interest in showing but who enjoy the continuing education with their companion dogs. Whatever their reason for studying with Marsha, they all benefit from her distinctive method of helping them polish their performance while fine tuning their communication with their dogs. Many of her students continue working with Marsha for years—such as the student who has been coming to classes for more than twenty years accompanied by twelve different dogs or the student and her dog who went from Novice A to OTCH with Marsha's coaching.

Marsha has always enjoyed the rewards of working with the more challenging breeds as well as those breeds that are more prevalent in the obedience ring. She began her years of participation in obedience trials competing with her two Scottish Terriers; Reed's Bawdy Becca Bairn UD and Reed's Highland Heather UD. Other partners through the years have been; Shetland Sheepdogs Jason's Miss Pollyanna UD and Starhill's Sweet Triathlete CDX (who passed away suddenly at two and a half years of age having already earned two legs of his UD), Rottweiler Dana's Oliver CDX, and Chinese Crested Expression–Mr. Mojo Working UDX CGC TDI (Dante).

Marsha's Chinese Crested partner, Jack (OTCH Przemek's Mr. Mojo Risin UDX CGC TDI), is the first of his breed to earn an AKC Obedience Trial Championship. Marsha and Jack have competed at the Gaines Eastern Regional Dog Obedience Championships for several years, at every level from Novice to Super Dog—placing 4th in Utility in 1997. Jack has been invited to compete in the Toy Group at the AKC

National Obedience Invitational four times—winning a Silver Medal in 2000 and a Gold Medal in 2001. He was a finalist for Therapy Dog of the Year in 2003.

Marsha and her dogs have participated in Agility, Tracking, Rally, and Canine Freestyle—and for several years, she and her Chinese Cresteds, Jack and Dante, have visited weekly with the residents at a nearby nursing home.

Marsha's website: www.canineworkout.com

SHALINI BOSBYSHELL

Shalini has lived with animals all her life and has had the ability to communicate with them since childhood. She and her husband, Hal, live with a large, ever-evolving, multi-species family of dogs, cats, horses, and many others. Animals have always been central to Shalini's life and work. She has extensive professional and personal experience with horses and has worked closely with veterinarians for many years.

Shalini accompanies her canine family members to dog school, and has participated in obedience and lure coursing trials and therapy dog work. Some of her training partners: Great Dane Mariway Bhima CGC TDI (who never competed, but trained through Utility), and Scottish Deerhounds Darkwynd Mushika CD, Highstone Periodos D'Lux JC, and CH Vale View December CGC TDI.

Her academic training is in physics, mathematics, geology, computer science, and Sanskrit language and literature. She has been devoted to the practice of meditation and the study of yoga philosophy for over thirty years.

Shalini has been offering private interspecies communication consultations and instruction for several years and her service is regularly recommended by veterinarians and trainers. Her work and education in the field of personal coaching have given her insight into the themes of learning and personal growth as well as the dynamics of interpersonal relationships.

It is her desire to help others discover the connection that exists between beings of all species, as greater awareness of this connection often can lead to spiritual growth and to a deeper understanding of the needs and desires of others.

Shalini's website: www.shalinibosbyshell.com